Elements of
Linear and
Multilinear
Algebra

Elements of

Linear and Multilinear Algebra

John M. Erdman
Portland State University, USA

 **World Scientific**

NEW JERSEY · LONDON · SINGAPORE · BEIJING · SHANGHAI · HONG KONG · TAIPEI · CHENNAI · TOKYO

Published by

World Scientific Publishing Co. Pte. Ltd.

5 Toh Tuck Link, Singapore 596224

USA office: 27 Warren Street, Suite 401-402, Hackensack, NJ 07601

UK office: 57 Shelton Street, Covent Garden, London WC2H 9HE

Library of Congress Cataloging-in-Publication Data

Names: Erdman, John M., 1935–   author.

Title: Elements of linear and multilinear algebra / John M. Erdman, Portland State University, USA.

Description: New Jersey : World Scientific, [2021] | Includes bibliographical references and index.

Identifiers: LCCN 2020052518 (print) | LCCN 2020052519 (ebook) |
  ISBN 9789811222726 (hardcover) | ISBN 9789811222733 (ebook for institutions) |
  ISBN 9789811222740 (ebook for individuals)

Subjects: LCSH: Algebras, Linear. | Multilinear algebra.

Classification: LCC QA184.2 .E73 2021 (print) | LCC QA184.2 (ebook) | DDC 512/.5--dc23

LC record available at https://lccn.loc.gov/2020052518

LC ebook record available at https://lccn.loc.gov/2020052519

British Library Cataloguing-in-Publication Data

A catalogue record for this book is available from the British Library.

For any available supplementary material, please visit
https://www.worldscientific.com/worldscibooks/10.1142/11896#t=suppl

Desk Editor: Soh Jing Wen

Typeset by Stallion Press
Email: enquiries@stallionpress.com

Printed in Singapore

It is not essential for the value of an education that every idea be understood at the time of its accession. Any person with a genuine intellectual interest and a wealth of intellectual content acquires much that he only gradually comes to understand fully in the light of its correlation with other related ideas... Scholarship is a progressive process, and it is the art of so connecting and recombining individual items of learning by the force of one's whole character and experience that nothing is left in isolation, and each idea becomes a commentary on many others.

— NORBERT WIENER

CONTENTS

PREFACE

This set of notes is an activity-oriented introduction to linear and multi-linear algebra. The great majority of the most elementary results in these subjects are straightforward and can be verified by the thoughtful student. Indeed, that is the main point of these notes — to convince the beginner that the subject is accessible. In the material that follows there are numerous indicators that suggest activity on the part of the reader: words such as "proposition", "example", "theorem", "exercise", and "corollary", if not followed by a proof (and proofs here are *very* rare) or a reference to a proof, are invitations to verify the assertions made. When the proof of a result appears to me to be too difficult for the average student to (re)invent and I have no improvements to offer to the standard proofs, I provide references to standard treatments.

These notes are intended to accompany an (academic) year-long course at the advanced undergraduate or beginning graduate level. (With judicious pruning most of the material can be covered in a two-term sequence.) When I taught the course I used a (rather highly modified) Moore-method. There were no lectures, but students were permitted to consult outside sources. Proofs of results were given, and discussed at considerable length, by students during class. The text is also suitable for a lecture-style class, the instructor proving some of the results while leaving others as exercises for the students.

The prerequisites for working through this material are quite modest. Elementary properties of the real number system, the arithmetic of matrices, ability to solve systems of linear equations, and the ability to evaluate the determinant of a square matrix are assumed. Although not logically required, an elementary course in calculus should help enormously in under-standing the point of the material in Chapters 6 and 9–12.

I have tried to keep the facts about vector spaces and those about inner product spaces separate. Many beginning linear algebra texts conflate the material on these two vastly different subjects. They differ in both notation and terminology. Terms such as "direct sum" and "projection" mean one thing in one context and something quite different in the other. (*E.g.*, consider the question, "Is the Euclidean plane the direct sum of the x-axis and the line $y = x$?" In vector spaces the answer is *yes*; in inner product spaces, *no*.) Inner product spaces have a much richer structure than vector spaces; they make possible the consideration of geometric and topological questions, and words like "length", "angle", "distance", and "perpendicular" make sense.

Vector spaces are the subject of Chapters 1, 2, and 4. Chapter 3 provides, for future convenience, a brief interlude on the language of categories for those not already familiar with it. There is no 'category theory' here, just the language. Chapter 5 concludes the material on linear algebra with a discussion of inner product spaces.

It seems to me that the extreme compartmentalization of mathematical education is not an entirely unmixed blessing. While it is surely true that one cannot ask to be well-educated in all fields of a subject so mind-numbingly enormous as mathematics, it is arguably desirable, when attempting to learn about highly abstract mathematical objects, that one be provided with some guidance enabling one to appreciate the new concept's importance in fields with which one is already familiar. This presents something of a dilemma in teaching multilinear algebra. Most treatments of multilinear algebra occur in algebra books. And algebra books (for good reasons) stick mainly to algebra. But the number of concepts arising in multilinear algebra is staggering. And the examples given in most texts are as abstract and unfamiliar to the beginning student as the concepts themselves. On the other hand, every student, I suppose, who has reached the level of multilinear algebra is already familiar with calculus. And the proper understanding of calculus requires an understanding of multilinearity. So, in this text, intended to be introductory, rather than rapidly summarize a whole herd of different subfields of multilinear algebra, I highlight one structure: Grassmann (or exterior) algebras and one example from calculus: differential forms.

One may reasonably wonder if the reformulation of the familiar calculus in terms of differential forms on manifolds is worth the effort. I think the answer is *yes* — it provides clarity. Over many years of teaching, I have found a few students who, having had only the standard presentation of calculus, can even remember what *the divergence theorem, the fundamental theorem of calculus, Green's theorem, Stokes' theorem,* and *the fundmental theorem for line integrals* say. And *none* who can explain why they are all exactly the same theorem, except in different dimensions. Why this is so, is made clear by the *Generalized Stokes' Theorem.* (See, in particular, Equation (12.4) and the exercises that follow it.)

Chapter 6 is a brief review of elementary calculus making use of previous material in the text to develop an appropriate coordinate-free definition of the *differential.* Chapters 7 and 8 introduce multilinear maps, determinants, tensor products, and Grassmann algebras. Chapters 9 and 10 are devoted to an important example of Grassmann algebras: the algebra of differential forms on a manifold.

Chapter 11 introduces just enough elementary homology theory so that boundary operators can be meaningfully discussed. This makes a possible discussion of *Stokes' theorem* in Chapter 12, which beautifully displays the enormous power of multilinear algebra in the familiar field of calculus.

In Chapter 13 we turn to yet another application of multilinear algebra to a familiar subject, plane geometry. We define a new multiplication on a particular Grassmann algebra that produces a novel approach to this subject that keeps track of the orientation (as well as similarity and congruence) of plane figures. This turns out to be an example of something called a *Clifford algebra.* These algebras have recently become hugely important in many fields of engineering, robotics, and theoretical physics. A full year's course would barely scratch the surface of what is now known about *Clifford algebras.* In Chapter 14 we look at a few very basic properties of these algebras and provide a basket of examples to illustrate the enormous variety of familiar things that looked at properly "are" *Clifford algebras.*

Concerning the materials in these notes, I make no claims of originality. There will surely be errors. I will be delighted to receive corrections, suggestions, or criticism at

erdmanj@comcast.net

NOTATION AND TERMINOLOGY

Some Algebraic Objects

Let S be a nonempty set. Consider the following axioms:

(1) $+\colon S \times S \to S$. ($+$ is a *binary operation*, called *addition*, on S)

(2) $(x+y)+z = x+(y+z)$ for all x, y, $z \in S$. (*associativity* of addition)

(3) There exists $\mathbf{0}_S \in S$ such that $x + \mathbf{0}_S = \mathbf{0}_S + x = x$ for all $x \in S$. (existence of an *additive identity*)

(4) For every $x \in S$ there exists $-x \in S$ such that $x + (-x) = (-x) + x = \mathbf{0}_S$. (existence of *additive inverses*)

(5) $x + y = y + x$ for all x, $y \in S$. (*commutativity* of addition)

(6) $\cdot\colon S \times S \to S\colon (x,y) \mapsto x \cdot y$. (the map $(x,y) \mapsto x \cdot y$ is a *binary operation*, called *multiplication*, on S. *Convention:* We will usually write xy instead of $x \cdot y$.)

(7) $(xy)z = x(yz)$ for all x, y, $z \in S$. (*associativity* of multiplication)

(8) $(x + y)z = xz + yz$ and $x(y + z) = xy + xz$ for all x, y, $z \in S$. (*multiplication distributes over addition*)

(9) There exists $\mathbf{1}_S$ in S such that $x\,\mathbf{1}_S = \mathbf{1}_S\,x = x$ for all $x \in S$. (existence of a *multiplicative identity* or unit)

(10) $\mathbf{1}_S \neq \mathbf{0}_S$.

(11) For every $x \in S$ such that $x \neq \mathbf{0}_S$ there exists $x^{-1} \in S$ such that $xx^{-1} = x^{-1}x = \mathbf{1}_S$. (existence of *multiplicative inverses*)

(12) $xy = yx$ for all x, $y \in S$. (*commutativity* of multiplication)

Definitions

- $(S, +)$ is a SEMIGROUP if it satisfies axioms (1)–(2).
- $(S, +)$ is a MONOID if it satisfies axioms (1)–(3).
- $(S, +)$ is a GROUP if it satisfies axioms (1)–(4).
- $(S, +)$ is an ABELIAN GROUP if it satisfies axioms (1)–(5).
- $(S, +, m)$ is a RING if it satisfies axioms (1)–(8).
- $(S, +, m)$ is a COMMUTATIVE RING if it satisfies axioms (1)–(8) and (12).

- $(S, +, m)$ is a UNITAL RING (or RING WITH IDENTITY, or UNITARY RING) if it satisfies axioms (1)–(9).
- $(S, +, m)$ is a DIVISION RING (or SKEW FIELD) if it satisfies axioms (1)–(11).
- $(S, +, m)$ is a FIELD if it satisfies axioms (1)–(12).

Remarks

- A binary operation is often written additively, $(x, y) \mapsto x + y$, if it is commutative and multiplicatively, $(x, y) \mapsto xy$, if it is not. This is by no means always the case: in a commutative ring (the real numbers or the complex numbers, for example), both addition and multiplication are commutative.
- When no confusion is likely to result we often write $\mathbf{0}$ for $\mathbf{0}_S$ and $\mathbf{1}$ for $\mathbf{1}_S$.
- Many authors require a *ring* to satisfy axioms (1)–(9).
- It is easy to see that axiom (10) holds in any unital ring except the *trivial ring* $S = \{\mathbf{0}\}$. *Convention:* Unless the contrary is stated we will assume that every unital ring is nontrivial.

Notation for Sets of Numbers

Here is a list of fairly standard notations for some sets of numbers which occur frequently in these notes:

$\mathbb{C}$ is the set of complex numbers

$\mathbb{R}$ is the set of real numbers

$\mathbb{R}^n$ is the set of all n-tuples $(r_1, r_2, \ldots, r_n)$ of real numbers

$\mathbb{R}^+ = \{x \in \mathbb{R} : x \geq 0\}$, the positive real numbers

$\mathbb{Q}$ is the set of rational numbers

$\mathbb{Q}^+ = \{x \in \mathbb{Q} : x \geq 0\}$, the positive rational numbers

$\mathbb{Z}$ is the set of integers

$\mathbb{Z}^+ = \{0, 1, 2, \dots\}$, the positive integers

$\mathbb{N} = \{1, 2, 3, \dots\}$, the set of natural numbers

$\mathbb{N}_n = \{1, 2, 3, \dots, n\}$ the first n natural numbers

$[a, b] = \{x \in \mathbb{R} : a \leq x \leq b\}$

$[a, b) = \{x \in \mathbb{R} : a \leq x < b\}$

$(a, b] = \{x \in \mathbb{R} : a < x \leq b\}$

$(a, b) = \{x \in \mathbb{R} : a < x < b\}$

$[a, \infty) = \{x \in \mathbb{R} : a \leq x\}$

$(a, \infty) = \{x \in \mathbb{R} : a < x\}$

$(-\infty, b] = \{x \in \mathbb{R} : x \leq b\}$

$(-\infty, b) = \{x \in \mathbb{R} : x < b\}$

$\mathbb{S}^1 = \{(x, y) \in \mathbb{R}^2 : x^2 + y^2 = 1\}$, the unit circle

Greek Letters

Upper case	Lower case	English name (approximate pronunciation)
A	α	Alpha (AL-fuh)
B	β	Beta (BAY-tuh)
Γ	γ	Gamma (GAM-uh)
Δ	δ	Delta (DEL-tuh)
E	ϵ or ε	Epsilon (EPP-suh-lon)
Z	ζ	Zeta (ZAY-tuh)
H	η	Eta (AY-tuh)
Θ	θ	Theta (THAY-tuh)
I	ι	Iota (eye-OH-tuh)
K	κ	Kappa (KAP-uh)
Λ	λ	Lambda (LAM-duh)
M	μ	Mu (MYOO)
N	ν	Nu (NOO)
Ξ	ξ	Xi (KSEE)
O	o	Omicron (OHM-ih-kron)
Π	π	Pi (PIE)
P	ρ	Rho (ROH)
Σ	σ	Sigma (SIG-muh)
T	τ	Tau (TAU)
Y	υ	Upsilon (OOP-suh-lon)
Φ	ϕ	Phi (FEE or FAHY)
X	χ	Chi (KHAY)
Ψ	ψ	Psi (PSEE or PSAHY)
Ω	ω	Omega (oh-MAY-guh)

Fraktur Fonts

In these notes Fraktur fonts are used (most often for families of sets and families of linear maps). Below are the Roman equivalents for each letter. When writing longhand or presenting material on a blackboard it is usually best to substitute script English letters.

Fraktur Upper case	Fraktur Lower case	Roman Lower Case
𝔄	𝔞	a
𝔅	𝔟	b
ℭ	𝔠	c
𝔇	𝔡	d
𝔈	𝔢	e
𝔉	𝔣	f
𝔊	𝔤	g
ℌ	𝔥	h
ℑ	𝔦	i
𝔍	𝔧	j
𝔎	𝔨	k
𝔏	𝔩	l
𝔐	𝔪	m
𝔑	𝔫	n
𝔒	𝔬	o
𝔓	𝔭	p
𝔔	𝔮	q
ℜ	𝔯	r
𝔖	𝔰	s
𝔗	𝔱	t
𝔘	𝔲	u
𝔙	𝔳	v
𝔚	𝔴	w
𝔛	𝔵	x
𝔜	𝔶	y
ℨ	𝔷	z

Chapter 1

VECTOR SPACES

1.1. Abelian Groups

Convention 1.1.1. In general, Abelian groups are written additively. That is, the usual notation for the binary operation on an Abelian group is +. Of course, there are special examples where this notation is inappropriate: most notably for the nonzero real numbers, the strictly positive real numbers, and the nonzero complex numbers under multiplication. It is conventional, if not entirely logical to write, "let G be an Abelian group," when what is meant is, "let $(G, +)$ be an Abelian group."

Proposition 1.1.2. *The identity element in an Abelian group is unique.*

Hint for proof. Let G be an Abelian group. Suppose that there are elements $\mathbf{0}$ and $\widetilde{\mathbf{0}}$ in G such that $x + \mathbf{0} = x$ and $x + \widetilde{\mathbf{0}} = x$ hold for every $x \in G$. Prove that $\mathbf{0} = \widetilde{\mathbf{0}}$.

Proposition 1.1.3. *Each element in an Abelian group has a unique inverse.*

Hint for proof. Let G be an Abelian group and $x \in G$. To prove uniqueness of the inverse for x suppose that there are elements y and z in G such that $x + y = \mathbf{0}$ and $x + z = \mathbf{0}$. Prove that $y = z$.

Proposition 1.1.4. *If x is an element of an Abelian group such that $x + x = x$, then $x = \mathbf{0}$.*

Proposition 1.1.5. *For every element x in an Abelian group $-(-x) = x$.*

Example 1.1.6. Let S be a nonempty set and $\mathbb{F}$ be a field. Denote by $\mathcal{F}(S, \mathbb{F})$ the family of all $\mathbb{F}$-valued functions on S. For $f, g \in \mathcal{F}(S, \mathbb{F})$ define $f + g$ by

$$(f + g)(x) = f(x) + g(x)$$

for all $x \in S$. Under this operation (called *pointwise addition*) $\mathcal{F}(S, \mathbb{F})$ is an Abelian group.

Example 1.1.7. As a special case of Example 1.1.6, we may regard Euclidean n-space $\mathbb{R}^n$ as an Abelian group.

Example 1.1.8. As a special case of Example 1.1.6, we may regard the set $\mathbb{R}^\infty$ of all sequences of real numbers as an Abelian group.

Example 1.1.9. Let $\mathbb{E}^2$ be the Euclidean plane. It contains points (which do *not* have coordinates) and lines (which do not have equations). A DIRECTED SEGMENT is an ordered pair of points. Define two directed segments to be *equivalent* if they are congruent (have the same length), lie on parallel lines, and have the same direction. This is clearly an equivalence relation on the set $\mathfrak{DS}$ of directed segments in the plane. We denote by $\overrightarrow{PQ}$ the equivalence class containing the directed segment (P, Q), going from the point P to the point Q. Define an operation $+$ on these equivalence classes by

$$\overrightarrow{PQ} + \overrightarrow{QR} = \overrightarrow{PR}.$$

This operation is well defined and under it $\mathfrak{DS}$ is an Abelian group.

Exercise 1.1.10. Suppose that A, B, C, and D are points in the plane such that $\overrightarrow{AB} = \overrightarrow{CD}$. Show that $\overrightarrow{AC} = \overrightarrow{BD}$.

Definition 1.1.11. Let G and H be Abelian groups. A map $f\colon G \to H$ is a HOMOMORPHISM if

$$f(x + y) = f(x) + f(y)$$

for all $x, y \in G$. We will denote by $\mathrm{Hom}(G, H)$ the set of all homomorphisms from G into H and will abbreviate $\mathrm{Hom}(G, G)$ to $\mathrm{Hom}(G)$.

Proposition 1.1.12. *If $f\colon G \to H$ is a homomorphism of Abelian groups, then $f(\mathbf{0}) = \mathbf{0}$.*

Proposition 1.1.13. *If $f\colon G \to H$ is a homomorphism of Abelian groups, then $f(-x) = -f(x)$ for each $x \in G$.*

Definition 1.1.14. Let G and H be Abelian groups. For f and g in $\mathrm{Hom}(G, H)$ we define

$$f + g\colon G \to H\colon x \mapsto f(x) + g(x).$$

Example 1.1.15. Let G and H be Abelian groups. With addition as defined in 1.1.14 $\mathrm{Hom}(G, H)$ is an Abelian group.

Hint for proof. Don't forget to show that $f + g$ belongs to $\mathrm{Hom}(G, H)$ when f and g do.

Convention 1.1.16. Let G, H, and J be Abelian groups and $f\colon G \to H$ and $g\colon H \to J$ be homomorphisms. Then the composite of g with f is denoted by gf (rather than by $g \circ f$). That is,

$$gf\colon G \to J\colon x \mapsto g(f(x)).$$

Proposition 1.1.17. *Let G, H, and J be Abelian groups, $f \in \mathrm{Hom}(G, H)$, and $g \in \mathrm{Hom}(H, J)$, then the composite gf belongs to $\mathrm{Hom}(G, J)$.*

1.2. Functions and Diagrams

Definition 1.2.1. Let S and T be sets and $f\colon S \to T$. The set S is the DOMAIN of f. The set T is the CODOMAIN of f, while $\{f(x)\colon x \in S\}$ is its RANGE. And $\{(x, f(x))\colon x \in S\}$ is the GRAPH of f. The domain of f is denoted by $\mathrm{dom}\, f$ and its range by $\mathrm{ran}\, f$.

Definition 1.2.2. A function f is INJECTIVE (or ONE-TO-ONE) if $x = y$ whenever $f(x) = f(y)$. That is, f is injective if no two distinct elements in its domain have the same image. An injective map is called an INJECTION.

A function is SURJECTIVE (or ONTO) if its range is equal to its codomain. A surjective map is called a SURJECTION.

A function is BIJECTIVE (or a ONE-TO-ONE CORRESPONDENCE) if it is both injective and surjective. A bijective map is called a BIJECTION.

Definition 1.2.3. It is frequently useful to think of functions as arrows in diagrams. For example, the situation $h\colon R \to S$, $j\colon R \to T$, $k\colon S \to U$,

$f\colon T \to U$ may be represented by the following diagram.

The diagram is said to COMMUTE if $k \circ h = f \circ j$. Diagrams need not be rectangular. For instance,

is a commutative diagram if $d = k \circ h$.

Example 1.2.4. Here is one diagrammatic way of stating the associative law for composition of functions: If $h\colon R \to S$, $g\colon S \to T$, and $f\colon T \to U$ and we define j and k so that the triangles in the diagram

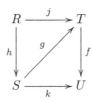

commute, then the square also commutes.

Convention 1.2.5. If S, T, and U are sets we will often not distinguish between $(S \times T) \times U$, $S \times (T \times U)$, and $S \times T \times U$. That is, the ordered pairs $\big((s,t),u\big)$ and $\big(s,(t,u)\big)$ and the ordered triple (s,t,u) will usually be treated as identical.

Notation 1.2.6. Let S be a set. The map

$$\mathrm{id}_S \colon S \to S \colon x \mapsto x$$

is the IDENTITY FUNCTION on S. When no confusion will result we write id for id_S.

Definition 1.2.7. Let S and T be sets, $f\colon S \to T$, and $A \subseteq S$. Then the RESTRICTION of f to A, denoted by $f\big|_A$, is the function $f \circ \iota_{A,S}$, where $\iota_{A,S}\colon A \to S\colon x \mapsto x$ is the *inclusion map* of A into S. That is, $f\big|_A$ is the mapping from A into T whose value at each x in A is $f(x)$.

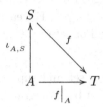

Suppose that $g\colon A \to T$ and $A \subseteq S$. A function $f\colon S \to T$ is an EXTENSION of g to S if $f\big|_A = g$, that is, if the diagram

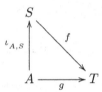

commutes.

Notation 1.2.8. If S, T, and U are nonempty sets and if $f\colon S \to T$ and $g\colon S \to U$, then we define the function $(f,g)\colon S \to T \times U$ by

$$(f,g)(s) = (f(s), g(s)).$$

Suppose, on the other hand, that we are given a function h mapping S into the Cartesian product $T \times U$. Then for each $s \in S$ the image $h(s)$ is an ordered pair, which we will write as $\big(h^1(s), h^2(s)\big)$. (The superscripts have nothing to do with powers.) Notice that we now have functions $h^1\colon S \to T$ and $h^2\colon S \to U$. These are the COMPONENTS of h. In abbreviated notation $h = (h^1, h^2)$.

Notation 1.2.9. Let $f\colon S \to U$ and $g\colon T \to V$ be functions between sets. Then $f \times g$ denotes the map

$$f \times g\colon S \times T \to U \times V\colon (s,t) \mapsto \big(f(s), g(t)\big).$$

Exercise 1.2.10. Let S be a set and $a\colon S \times S \to S$ be a function such that the diagram

$$S \times S \times S \underset{\mathrm{id} \times a}{\overset{a \times \mathrm{id}}{\rightrightarrows}} S \times S \overset{a}{\longrightarrow} S \tag{D1}$$

commutes. What is (S, a)? *Hint.* Interpret a as, for example, addition (or multiplication).

Convention 1.2.11. We will have use for a standard one-element set, which, if we wish, we can regard as the Cartesian product of an empty family of sets. We will denote it by **1**. For each set S there is exactly one function from S into **1**. We will denote it by ε_S. If no confusion is likely to arise we write ε for ε_S.

Exercise 1.2.12. Let S be a set and suppose that $a\colon S \times S \to S$ and $\eta\colon \mathbf{1} \to S$ are functions such that both diagram (D1) above and the diagram (D2) that follows commute.

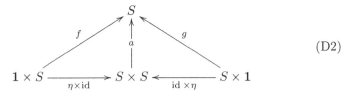

(D2)

(Here f and g are the obvious bijections.) What is (S, a, η)?

Notation 1.2.13. We denote by δ the *diagonal* mapping of a set S into $S \times S$. That is,

$$\delta\colon S \to S \times S\colon s \mapsto (s, s).$$

Exercise 1.2.14. Let S be a set and suppose that $a\colon S \times S \to S$ and $\eta\colon \mathbf{1} \to S$ are functions such that the diagrams (D1) and (D2) above commute. Suppose further that there is a function $\iota\colon S \to S$ for which the following diagram commutes.

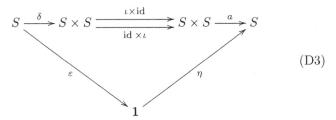

(D3)

What is (S, a, η, ι)?

Notation 1.2.15. Let S be a set. We denote by σ the INTERCHANGE (or SWITCHING) operation on the $S \times S$. That is,

$$\sigma\colon S \times S \to S \times S\colon (s, t) \mapsto (t, s).$$

Exercise 1.2.16. Let S be a set and suppose that $a\colon S \times S \to S$, $\eta\colon \mathbf{1} \to S$, and $\iota\colon S \to S$ are functions such that the diagrams (D1), (D2), and (D3) above commute. Suppose further that the following diagram commutes.

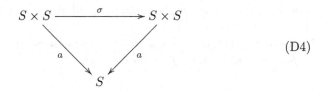

(D4)

What is $(S, a, \eta, \iota, \sigma)$?

Exercise 1.2.17. Let $f\colon G \to H$ be a function between Abelian groups. Suppose that the diagram

$$
\begin{array}{ccc}
G \times G & \xrightarrow{\ f \times f\ } & H \times H \\
\downarrow{\scriptstyle +} & & \downarrow{\scriptstyle +} \\
G & \xrightarrow{\ \ f\ \ } & H
\end{array}
$$

commutes. What can be said about the function f?

Notation 1.2.18. If S and T are sets we denote by $\mathcal{F}(S, T)$ the family of all functions from S into T. When $\mathbb{F}$ is a field there are several common notations for the family of $\mathbb{F}$-valued functions on S. We denote by $l(S)$ (or by $l(S, \mathbb{F})$, or by $\mathbb{F}^S$, or by $\mathcal{F}(S, \mathbb{F})$, or by $\mathcal{F}(S)$) the family of all functions $\alpha\colon S \to \mathbb{F}$. For $x \in l(S)$ we frequently write the value of x at $s \in S$ as x_s rather than $x(s)$. (Sometimes it seems a good idea to reduce the number of parentheses cluttering a page.)

The SUPPORT of a function $f\colon S \to \mathbb{F}$, denoted by $\mathrm{supp}(f)$, is $\{x \in S\colon f(x) \neq 0\}$.

Furthermore, we will denote by $l_c(S)$ (or by $l_c(S, \mathbb{F})$, or by $\mathcal{F}_c(S)$) the family of all functions $\alpha\colon S \to \mathbb{F}$ with finite support; that is, those functions on S that are nonzero at only finitely many elements of S.

Exercise 1.2.19. Let S be a set with exactly one element. Discuss the cardinality of (that is, the number of elements in) the sets $\mathcal{F}(\emptyset, \emptyset)$, $\mathcal{F}(\emptyset, S)$, $\mathcal{F}(S, \emptyset)$, and $\mathcal{F}(S, S)$,

1.3. Rings

Recall that an ordered triple $(R, +, \cdot)$ is a *ring* if $(R, +)$ is an Abelian group, $(R, \cdot)$ is a semigroup, and the distributive laws (see *Some Algebraic Objects* 8) hold. The ring is *unital* if, in addition, $(R \setminus \{\mathbf{0}\}, \cdot)$ is a monoid.

Proposition 1.3.1. *The additive identity of a ring is an annihilator. That is, for every element a of a ring $\mathbf{0}a = a\mathbf{0} = \mathbf{0}$.*

Proposition 1.3.2. *If a and b are elements of a ring, then $(-a)b = a(-b) = -(ab)$ and $(-a)(-b) = ab$.*

Proposition 1.3.3. *Let a and b be elements of a unital ring. Then $1 - ab$ is invertible if and only if $1 - ba$ is.*

Hint for proof. Look at the product of $1 - ba$ and $1 + bca$ where c is the inverse of $1 - ab$.

Definition 1.3.4. An element a of a ring is LEFT CANCELLABLE if $ab = ac$ implies that $b = c$. It is RIGHT CANCELLABLE if $ba = ca$ implies that $b = c$. A ring has the CANCELLATION PROPERTY if every nonzero element of the ring is both left and right cancellable.

Exercise 1.3.5. Every division ring has the cancellation property.

Definition 1.3.6. A nonzero element a of a ring is a ZERO DIVISOR (or DIVISOR OF ZERO) if there exists a nonzero element b of the ring such that (i) $ab = \mathbf{0}$ or (ii) $ba = \mathbf{0}$.

   Most everyone agrees that a nonzero element a of a ring is a *left divisor of zero* if it satisfies (i) for some nonzero b and a *right divisor of zero* if it satisfies (ii) for some nonzero b. There agreement on terminology ceases. Some authors ([6], for example) use the definition above for *divisor of zero*; others ([18], for example) require a *divisor of zero* to be *both* a left and a right divisor of zero; and yet others ([19], for example) avoid the issue entirely by defining *zero divisors* only for commutative rings. Palmer in [27] makes the most systematic distinctions: a *zero divisor* is defined as above; an element that is both a left and a right zero divisor is a *two-sided zero divisor*; and if the same nonzero b makes both (i) and (ii) hold a is a *joint zero divisor*.

Proposition 1.3.7. *A division ring has no zero divisors. That is, if $ab = 0$ in a division ring, then $a = 0$ or $b = 0$.*

Proposition 1.3.8. *A ring has the cancellation property if and only if it has no zero divisors.*

Example 1.3.9. Let G be an Abelian group. Then $\mathrm{Hom}(G)$ is a unital ring (under the operations of addition and composition).

Definition 1.3.10. A function $f \colon R \to S$ between rings is a (RING) HOMOMORPHISM if

$$f(x + y) = f(x) + f(y) \tag{1.1}$$

and

$$f(xy) = f(x)f(y) \tag{1.2}$$

for all x and y in R. If in addition R and S are unital rings and

$$f(\mathbf{1}_R) = \mathbf{1}_S \tag{1.3}$$

then f is a UNITAL (RING) HOMOMORPHISM.

Obviously a ring homomorphism $f \colon R \to S$ is a group homomorphism of R and S regarded as Abelian groups. The KERNEL of f as a ring homomorphism is the kernel of f as a homomorphism of Abelian groups; that is $\ker f = \{x \in R \colon f(x) = \mathbf{0}\}$.

If f^{-1} exists and is also a ring homomorphism, then f is an ISOMORPHISM from R to S. If an isomorphism from R to S exists, then R and S are ISOMORPHIC.

1.4. Vector Spaces

Definition 1.4.1. Let $\mathbb{F}$ be a field. An ordered triple $(V, +, M)$ is a VECTOR SPACE OVER $\mathbb{F}$ if $(V, +)$ is an Abelian group and $M \colon \mathbb{F} \to \mathrm{Hom}(V)$ is a unital ring homomorphism. An element of V is a VECTOR and an element of $\mathbb{F}$ is a SCALAR. A vector space whose scalars are real numbers is a REAL VECTOR SPACE and one with complex numbers as scalars is a COMPLEX VECTOR SPACE. The vector space $\{\mathbf{0}\}$ containing a single element is the TRIVIAL VECTOR SPACE.

Exercise 1.4.2. The definition of *vector space* found in many elementary texts is something like the following: a *vector space* (over a field $\mathbb{F}$) is a set V together with operations of addition and scalar multiplication that satisfy the following axioms:

(1) if $\mathbf{x}, \mathbf{y} \in V$, then $\mathbf{x} + \mathbf{y} \in V$;

(2) $(\mathbf{x} + \mathbf{y}) + \mathbf{z} = \mathbf{x} + (\mathbf{y} + \mathbf{z})$ for every $\mathbf{x}, \mathbf{y}, \mathbf{z} \in V$ (associativity);

(3) there exists $\mathbf{0} \in V$ such that $\mathbf{x} + \mathbf{0} = \mathbf{x}$ for every $\mathbf{x} \in V$ (existence of additive identity);

(4) for every $\mathbf{x} \in V$ there exists $-\mathbf{x} \in V$ such that $\mathbf{x} + (-\mathbf{x}) = \mathbf{0}$ (existence of additive inverses);

(5) $\mathbf{x} + \mathbf{y} = \mathbf{y} + \mathbf{x}$ for every $\mathbf{x}, \mathbf{y} \in V$ (commutativity);

(6) if $\alpha \in \mathbb{F}$ and $\mathbf{x} \in V$, then $\alpha \mathbf{x} \in V$;

(7) $\alpha(\mathbf{x} + \mathbf{y}) = \alpha \mathbf{x} + \alpha \mathbf{y}$ for every $\alpha \in \mathbb{F}$ and every $\mathbf{x}, \mathbf{y} \in V$;

(8) $(\alpha + \beta)\mathbf{x} = \alpha \mathbf{x} + \beta \mathbf{x}$ for every $\alpha, \beta \in \mathbb{F}$ and every $\mathbf{x} \in V$;

(9) $(\alpha \beta)\mathbf{x} = \alpha(\beta \mathbf{x})$ for every $\alpha, \beta \in \mathbb{F}$ and every $\mathbf{x} \in V$; and

(10) $1\mathbf{x} = \mathbf{x}$ for every $\mathbf{x} \in V$.

Verify that this definition is equivalent to the one given above in 1.4.1.

Proposition 1.4.3. *If x is an element of a vector space, then $(-1)x$ is the additive inverse of x. That is, $(-1)x = -x$. (Here, of course, 1 is the multiplicative identity of the field $\mathbb{F}$.)*

Example 1.4.4. Let $\mathbb{F}$ be a field. Then $\mathbb{F}$ can be regarded as a vector space over itself.

Example 1.4.5. Let S be a nonempty set and $\mathbb{F}$ be a field. In Example 1.1.6 we saw that the family $\mathcal{F}(S, \mathbb{F})$ of $\mathbb{F}$-valued functions on S is an Abelian group under pointwise addition. For $f \in \mathcal{F}(S, \mathbb{F})$ and $\alpha \in \mathbb{F}$ define αf by

$$(\alpha f)(x) = \alpha \cdot f(x)$$

for all $x \in S$. Under this operation (called *pointwise scalar multiplication*) the Abelian group $\mathcal{F}(S, \mathbb{F})$ becomes a vector space. When $\mathbb{F} = \mathbb{R}$ we write $\mathcal{F}(S)$ for $\mathcal{F}(S, \mathbb{R})$.

Example 1.4.6. As a special case of Example 1.4.5, we may regard Euclidean n-space $\mathbb{R}^n$ as a vector space.

Example 1.4.7. As another special case of Example 1.4.5, we may regard the set $\mathbb{R}^\infty$ of all sequences of real numbers as a vector space.

Example 1.4.8. Yet another special case of Example 1.4.5, is the vector space $\mathbf{M}_{m \times n}(\mathbb{F})$ of $m \times n$ matrices of members of $\mathbb{F}$ (where m, $n \in \mathbb{N}$). We will use $\mathbf{M}_n(\mathbb{F})$ as shorthand for $\mathbf{M}_{n \times n}(\mathbb{F})$ and $\mathbf{M}_n$ for $\mathbf{M}_n(\mathbb{R})$.

Exercise 1.4.9. Let V be the set of all real numbers. Define an operation of "addition" by

$$x \boxplus y = \text{the maximum of } x \text{ and } y$$

for all x, $y \in V$. Define an operation of "scalar multiplication" by

$$\alpha \boxdot x = \alpha x$$

for all $\alpha \in \mathbb{R}$ and $x \in V$. Prove or disprove: under the operations $\boxplus$ and $\boxdot$ the set V is a vector space.

Exercise 1.4.10. Let V be the set of all real numbers x such that $x > 0$. Define an operation of "addition" by

$$x \boxplus y = xy$$

for all x, $y \in V$. Define an operation of "scalar multiplication" by

$$\alpha \boxdot x = x^{\alpha}$$

for all $\alpha \in \mathbb{R}$ and $x \in V$. Prove or disprove: under the operations $\boxplus$ and $\boxdot$ the set V is a vector space.

Exercise 1.4.11. Let V be $\mathbb{R}^2$, the set of all ordered pairs (x, y) of real numbers. Define an operation of "addition" by

$$(u, v) \boxplus (x, y) = (u + x + 1, v + y + 1)$$

for all (u, v) and (x, y) in V. Define an operation of "scalar multiplication" by

$$\alpha \boxdot (x, y) = (\alpha x, \alpha y)$$

for all $\alpha \in \mathbb{R}$ and $(x, y) \in V$. Prove or disprove: under the operations $\boxplus$ and $\boxdot$ the set V is a vector space.

Exercise 1.4.12. Let V be the set of all $n \times n$ matrices of real numbers. Define an operation of "addition" by

$$A \boxplus B = \tfrac{1}{2}(AB + BA)$$

for all A, $B \in V$. Define an operation of "scalar multiplication" by

$$\alpha \boxdot A = \mathbf{0}$$

for all $\alpha \in \mathbb{R}$ and $A \in V$. Prove or disprove: under the operations $\boxplus$ and $\boxdot$ the set V is a vector space. (If you have forgotten how to multiply matrices, look in any beginning linear algebra text.)

Proposition 1.4.13. *If x is a vector and α is a scalar, then $\alpha x = \mathbf{0}$ if and only if $\alpha = 0$ or $x = \mathbf{0}$.*

In Example 1.1.9 we saw how to make the family of equivalence classes of directed segments in the plane into an Abelian group. We may also define scalar multiplication on these equivalence classes by declaring that

(1) if $\alpha > 0$, then $\alpha \overrightarrow{PQ} = \overrightarrow{PR}$ where P, Q, and R are collinear, P *does not* lie between Q and R, and the length of the directed segment (P, R) is α times the length of (P, Q);
(2) if $\alpha = 0$, then $\alpha \overrightarrow{PQ} = \overrightarrow{PP}$; and
(3) if $\alpha < 0$, then $\alpha \overrightarrow{PQ} = \overrightarrow{PR}$ where P, Q, and R are collinear, P *does* lie between Q and R, and the length of the directed segment (R, P) is α times the length of (P, Q).

Exercise 1.4.14. Show that the scalar multiplication presented above is well-defined and that it makes the Abelian group of equivalence classes of directed segments in the plane into a vector space.

Remark 1.4.15. Among the methods for proving elementary facts about Euclidean geometry of the plane three of the most common are *synthetic geometry*, *analytic geometry*, and *vector geometry*. In *synthetic geometry* points do not have coordinates, lines do not have equations, and vectors are not mentioned; but standard theorems from Euclid's *Elements* are used. *Analytic geometry* makes use of a coordinate system in terms of which points are assigned coordinates and lines are described by equations; little or no use is made of vectors or major theorems of Euclidean geometry. *Vector geometry* uses vectors as defined in the preceding exercise, but does not rely on Euclidean theorems or coordinate systems. Although there is nothing illogical about mixing these methods in establishing a result, it is interesting to try to construct separate proofs of some elementary results using each method in turn. That is what the next few exercises are about.

Exercise 1.4.16. Use each of the three geometric methods described above to show that the diagonals of a parallelogram bisect each other.

Exercise 1.4.17. Use each of the three geometric methods described above to show that if the diagonals of a quadrilateral bisect each other then the quadrilateral is a parallelogram.

Exercise 1.4.18. Use each of the three geometric methods described above to show that the line joining the midpoints of the nonparallel sides of a trapezoid is parallel to the bases and its length is half the sum of the lengths of the bases.

Exercise 1.4.19. Use each of the three geometric methods described above to show that the line segments joining the midpoints of adjacent sides of an arbitrary quadrilateral form a parallelogram.

Exercise 1.4.20. Three vertices of a parallelogram $PQRS$ in 3-space are $P = (1, 3, 2)$, $Q = (4, 5, 3)$, and $R = (2, -1, 0)$. What are the coordinates of the point S, opposite Q?

1.5. Subspaces

Definition 1.5.1. A subset M of a vector space V is a SUBSPACE of V if it is a vector space under the operations it inherits from V.

Notation 1.5.2. For a vector space V we will write $M \preceq V$ to indicate that M is a subspace of V. To distinguish this concept from other uses of the word "subspace" (*topological subspace*, for example) writers frequently use the expressions *linear subspace*, *vector subspace*, or *linear manifold*.

Proposition 1.5.3. *A nonempty subset of M of a vector space V is a subspace of V if and only if it is closed under addition and scalar multiplication. (That is: if $\mathbf{x}$ and $\mathbf{y}$ belong to M, so does $\mathbf{x} + \mathbf{y}$; and if $\mathbf{x}$ belongs to M and $\alpha \in \mathbb{F}$, then $\alpha\mathbf{x}$ belongs to M.*

Example 1.5.4. In each of the following cases prove or disprove that the set of points (x, y, z) in $\mathbb{R}^3$ satisfying the indicated condition is a subspace of $\mathbb{R}^3$.

(a) $x + 2y - 3z = 4$.
(b) $\dfrac{x-1}{2} = \dfrac{y+2}{3} = \dfrac{z}{4}$.
(c) $x + y + z = 0$ and $x - y + z = 1$.
(d) $x = -z$ and $x = z$.

(e) $x^2 + y^2 = z$.

(f) $\dfrac{x}{2} = \dfrac{y-3}{5}$.

Proposition 1.5.5. *Let $\mathcal{M}$ be a family of subspaces of a vector space V. Then the intersection $\bigcap \mathcal{M}$ of this family is itself a subspace of V.*

Exercise 1.5.6. Let A be a nonempty set of vectors in a vector space V. Explain carefully why it makes sense to say that the intersection of the family of all subspaces containing A is "the smallest subspace of V that contains A".

Exercise 1.5.7. Find and describe geometrically the smallest subspace of $\mathbb{R}^3$ containing the vectors $(0, -3, 6)$ and $(0, 1, -2)$.

Exercise 1.5.8. Find and describe geometrically the smallest subspace of $\mathbb{R}^3$ containing the vectors $(2, -3, -3)$ and $(0, 3, 2)$.

Example 1.5.9. Let $\mathbb{R}^\infty$ denote the vector space of all sequences of real numbers. (See Example 1.4.5.) In each of the following a subset of $\mathbb{R}^\infty$ is described. Prove or disprove that it is a subspace of $\mathbb{R}^\infty$.

(a) Sequences that have infinitely many zeros (for example, $(1, 1, 0, 1, 1, 0, 1, 1, 0, \ldots)$).
(b) Sequences that are eventually zero. (A sequence (x_k) is *eventually zero* if there is an index n_0 such that $x_n = 0$ whenever $n \geq n_0$.)
(c) Sequences that are absolutely summable. (A sequence (x_k) is *absolutely summable* if $\sum_{k=1}^\infty |x_k| < \infty$.)
(d) Bounded sequences. (A sequence (x_k) is *bounded* if there is a positive number M such that $|x_k| \leq M$ for every k.)
(e) Decreasing sequences. (A sequence (x_k) is *decreasing* if $x_{n+1} \leq x_n$ for each n.)
(f) Convergent sequences. (A sequence (x_k) is *convergent* if there is a number ℓ such that the sequence is eventually in every neighborhood of ℓ; that is, if there is a number ℓ such that for every $\epsilon > 0$ there exists $n_0 \in \mathbb{N}$ such that $|x_n - \ell| < \epsilon$ whenever $n \geq n_0$.)
(g) Arithmetic progressions. (A sequence (x_k) is *arithmetic* if it is of the form $(a, a + k, a + 2k, a + 3k, \ldots)$ for some constant k.)
(h) Geometric progressions. (A sequence (x_k) is *geometric* if it is of the form $(a, ka, k^2 a, k^3 a, \ldots)$ for some constant k.)

Notation 1.5.10. Here are some frequently encountered families of functions:

$\mathcal{F} = \mathcal{F}[a, b] = \{f \colon f \text{ is a real valued function on the interval } [a, b]\}$

$\mathcal{P} = \mathcal{P}[a, b] = \{p \colon p \text{ is a polynomial function on } [a, b]\}$

$\mathcal{P}_4 = \mathcal{P}_4[a, b] = \{p \in \mathcal{P} \colon \text{ the degree of } p \text{ is strictly less than 4}\}$

$\mathcal{Q}_4 = \mathcal{Q}_4[a, b] = \{p \in \mathcal{P} \colon \text{ the degree of } p \text{ is equal to 4}\}$

$\mathcal{C} = \mathcal{C}[a, b] = \{f \in \mathcal{F} \colon f \text{ is continuous}\}$

$\mathcal{D} = \mathcal{D}[a, b] = \{f \in \mathcal{F} \colon f \text{ is differentiable}\}$

$\mathcal{K} = \mathcal{K}[a, b] = \{f \in \mathcal{F} \colon f \text{ is a constant function}\}$

$\mathcal{B} = \mathcal{B}[a, b] = \{f \in \mathcal{F} \colon f \text{ is bounded}\}$

$\mathcal{J} = \mathcal{J}[a, b] = \{f \in \mathcal{F} \colon f \text{ is integrable}\}$

(A function $f \in \mathcal{F}$ is BOUNDED if there exists a number $M \geq 0$ such that $|f(x)| \leq M$ for all x in $[a, b]$. It is (RIEMANN) INTEGRABLE if it is bounded and $\int_a^b f(x)\,dx$ exists.)

Exercise 1.5.11. For a fixed interval $[a, b]$, which sets of functions in the list 1.5.10 are vector subspaces of which?

Notation 1.5.12. If A and B are subsets of a vector space then the SUM of A and B, denoted by $A + B$, is defined by

$$A + B := \{a + b \colon a \in A \text{ and } b \in B\}.$$

The set $A - B$ is defined similarly. For a set $\{a\}$ containing a single element we write $a + B$ instead of $\{a\} + B$.

Exercise 1.5.13. Let M and N be subspaces of a vector space V. Consider the following subsets of V.

(1) $M \cup N$. (A vector v belongs to $M \cup N$ if it belongs to either M or N.)
(2) $M + N$.
(3) $M \setminus N$. (A vector v belongs to $M \setminus N$ if it belongs to M but not to N.)
(4) $M - N$.

For each of the sets (a)–(d) above, either prove that it *is* a subspace of V or give a counterexample to show that it *need not* be a subspace of V.

Definition 1.5.14. Let M and N be subspaces of a vector space V. If $M \cap N = \{\mathbf{0}\}$ and $M + N = V$, then V is the (INTERNAL) DIRECT SUM of M and N. In this case we write

$$V = M \oplus N.$$

In this case the subspaces M and N are COMPLEMENTARY and each is the COMPLEMENT of the other.

Example 1.5.15. In $\mathbb{R}^3$ let M be the line $x = y = z$, N be the line $x = \frac{1}{2}y = \frac{1}{3}z$, and $L = M + N$. Then $L = M \oplus N$.

Example 1.5.16. Let M be the plane $x + y + z = 0$ and N be the line $x = y = z$ in $\mathbb{R}^3$. Then $\mathbb{R}^3 = M \oplus N$.

Example 1.5.17. Let $\mathcal{C} = \mathcal{C}[-1,1]$ be the vector space of all continuous real valued functions on the interval $[-1,1]$. A function f in $\mathcal{C}$ is EVEN if $f(-x) = f(x)$ for all $x \in [-1,1]$; it is ODD if $f(-x) = -f(x)$ for all $x \in [-1,1]$. Let $\mathcal{C}_o = \{f \in \mathcal{C} : f \text{ is odd }\}$ and $\mathcal{C}_e = \{f \in \mathcal{C} : f \text{ is even }\}$. Then $\mathcal{C} = \mathcal{C}_o \oplus \mathcal{C}_e$.

Example 1.5.18. Let $\mathcal{C} = \mathcal{C}[0,1]$ be the family of continuous real valued functions on the interval $[0,1]$. Define

$$f_1(t) = t \quad \text{and} \quad f_2(t) = t^4$$

for $0 \le t \le 1$. Let M be the set of all functions of the form $\alpha f_1 + \beta f_2$ where $\alpha, \beta \in \mathbb{R}$. And let N be the set of all functions g in $\mathcal{C}$ that satisfy

$$\int_0^1 t g(t)\, dt = 0 \quad \text{and} \quad \int_0^1 t^4 g(t)\, dt = 0.$$

Then $\mathcal{C} = M \oplus N$.

Exercise 1.5.19. In the preceding example let $g(t) = t^2$ for $0 \le t \le 1$. Find polynomials $f \in M$ and $h \in N$ such that $f = g + h$.

Theorem 1.5.20 (Vector Decomposition Theorem). *Let V be a vector space such that $V = M \oplus N$. Then for every vector $v \in V$ there exist unique vectors $m \in M$ and $n \in N$ such that $v = m + n$.*

Exercise 1.5.21. Define what it means for a vector space V to be the direct sum of subspaces $M_1, \ldots, M_n$. Show (using your definition) that if V is the direct sum of these subspaces, then for every $v \in V$ there exist unique vectors $m_k \in M_k$ (for $k = 1, \ldots, n$) such that $v = m_1 + \cdots + m_n$.

1.6. Linear Combinations and Linear Independence

Some authors of linear algebra texts make it appear as if the terms *linear dependence* and *linear independence*, *span*, and *basis* pertain only to *finite* sets of vectors. This is misleading. The terms should make sense for *arbitrary* sets. In particular, do not be misled into believing that a basis for a vector space must be a finite set of vectors (or a sequence of vectors).

Definition 1.6.1. A vector y is a LINEAR COMBINATION of distinct vectors $x_1, \ldots, x_n$ if there exist scalars $\alpha_1, \ldots \alpha_n$ such that $y = \sum_{k=1}^{n} \alpha_k x_k$. *Note:* linear combinations *are* finite sums. The linear combination $\sum_{k=1}^{n} \alpha_k \mathbf{x}_k$ is TRIVIAL if all the coefficients $\alpha_1, \ldots \alpha_n$ are zero. If at least one α_k is different from zero, the linear combination is NONTRIVIAL.

Example 1.6.2. In $\mathbb{R}^2$ the vector $(8, 2)$ is a linear combination of the vectors $(1, 1)$ and $(1, -1)$.

Example 1.6.3. In $\mathbb{R}^3$ the vector $(1, 2, 3)$ is *not* a linear combination of the vectors $(1, 1, 0)$ and $(1, -1, 0)$.

Definition 1.6.4. Let A be a subset of a vector space V. The SPAN of A is the intersection of the family of all subspaces of V that contain A. It is denoted by $\operatorname{span}(A)$ (or by $\operatorname{span}_{\mathbb{F}}(A)$ if we wish to emphasize the role of the scalar field $\mathbb{F}$). The subset A SPANS the space V if $V = \operatorname{span}(A)$. In this case we also say that A is a SPANNING SET for V.

Proposition 1.6.5. *If A is a nonempty subset of a vector space V, then* $\operatorname{span} A$ *is the set of all linear combinations of elements of A.*

Remark 1.6.6. Occasionally one must consider the not-too-interesting question of what is meant by the *span* of the empty set. According to the "abstract" Definition 1.6.4 above it is the intersection of all the subspaces that contain the empty set. That is, $\operatorname{span} \emptyset = \{\mathbf{0}\}$. (Had we preferred Proposition 1.6.5 as a more "constructive" definition of *span* — the set of all linear combinations of elements in $\emptyset$ — then the span of the empty set would have been just $\emptyset$ itself.)

Example 1.6.7. For each $n = 0, 1, 2, \ldots$ define a function p_n on $\mathbb{R}$ by $p_n(x) = x^n$. Let $\mathcal{P}$ be the set of polynomial functions on $\mathbb{R}$. It is a subspace of the vector space of continuous functions on $\mathbb{R}$. Then $\mathcal{P} = \operatorname{span}\{p_0, p_1, p_2, \ldots\}$. The exponential function exp, whose value at x is e^x, is not in the span of the set $\{p_0, p_1, p_2 \ldots\}$.

Definition 1.6.8. A subset A (finite or not) of a vector space is LINEARLY DEPENDENT if the zero vector $\mathbf{0}$ can be written as a nontrivial linear combination of elements of A; that is, if there exist distinct vectors $x_1, \ldots, x_n \in A$ and scalars $\alpha_1, \ldots, \alpha_n$, **not all zero**, such that $\sum_{k=1}^{n} \alpha_k x_k = \mathbf{0}$. A subset of a vector space is LINEARLY INDEPENDENT if it is not linearly dependent.

Technically, it is a *set* of vectors that is linearly dependent or independent. Nevertheless, these terms are frequently used as if they were properties of the vectors themselves. For instance, if $S = \{x_1, \ldots, x_n\}$ is a finite set of vectors in a vector space, you may see the assertions "the set S is linearly independent" and "the vectors $x_1, \ldots, x_n$ are linearly independent" used interchangeably.

Supersets of linearly dependent sets are linearly dependent and subsets of linearly independent sets linearly independent.

Proposition 1.6.9. *Suppose that V is a vector space and $A \subseteq B \subseteq V$. If A is linearly dependent, then so is B. Equivalently, if B is linearly independent, then so is A.*

Exercise 1.6.10. Let $w = (1, 1, 0, 0)$, $x = (1, 0, 1, 0)$, $y = (0, 0, 1, 1)$, and $z = (0, 1, 0, 1)$.

(a) Show that $\{w, x, y, z\}$ does not span $\mathbb{R}^4$ by finding a vector u in $\mathbb{R}^4$ such that $u \notin \text{span}(w, x, y, z)$.

(b) Show that $\{w, x, y, z\}$ is a linearly dependent set of vectors by finding scalars α, β, γ, and δ — not all zero — such that $\alpha w + \beta x + \gamma y + \delta z = 0$.

(c) Show that $\{w, x, y, z\}$ is a linearly dependent set by writing z as a linear combination of w, x, and y.

Example 1.6.11. The (vectors going from the origin to) points on the unit circle in $\mathbb{R}^2$ are linearly dependent.

Example 1.6.12. For each $n = 0, 1, 2, \ldots$ define a function p_n on $\mathbb{R}$ by $p_n(x) = x^n$. Then the set $\{p_0, p_1, p_2, \ldots\}$ is a linearly independent subset of the vector space of continuous functions on $\mathbb{R}$.

Example 1.6.13. In the vector space $\mathcal{C}[0, \pi]$ of continuous functions on the interval $[0, \pi]$ define the vectors f, g, and h by

$$f(x) = x$$
$$g(x) = \sin x$$
$$h(x) = \cos x$$

for $0 \leq x \leq \pi$. Then f, g, and h are linearly independent.

Example 1.6.14. In the vector space $\mathcal{C}[0, \pi]$ of continuous functions on $[0, \pi]$ let f, g, h, and j be the vectors defined by

$$f(x) = 1$$
$$g(x) = x$$
$$h(x) = \cos x$$
$$j(x) = \cos^2 \frac{x}{2}$$

for $0 \le x \le \pi$. Then f, g, h, and j are linearly dependent.

Exercise 1.6.15. Let a, b, and c be distinct real numbers. Show that the vectors $(1, 1, 1)$, (a, b, c), and (a^2, b^2, c^2) form a linearly independent subset of $\mathbb{R}^3$.

Exercise 1.6.16. In the vector space $\mathcal{C}[0, 1]$ define the vectors f, g, and h by

$$f(x) = x$$
$$g(x) = e^x$$
$$h(x) = e^{-x}$$

for $0 \le x \le 1$. Are f, g, and h linearly independent?

Exercise 1.6.17. Let $u = (\lambda, 1, 0)$, $v = (1, \lambda, 1)$, and $w = (0, 1, \lambda)$. Find **all** values of λ that make $\{u, v, w\}$ a linearly dependent subset of $\mathbb{R}^3$.

Exercise 1.6.18. Suppose that $\{u, v, w\}$ is a linearly independent set in a vector space V. Show that the set $\{u+v, u+w, v+w\}$ is linearly independent in V.

1.7. Bases for Vector Spaces

Definition 1.7.1. A set B (finite or not) of vectors in a vector space V is a BASIS for V if it is linearly independent and spans B.

Example 1.7.2. The vectors $e^1 = (1, 0, 0)$, $e^2 = (0, 1, 0)$, and $e^3 = (0, 0, 1)$ constitute an ordered basis for the vector space $\mathbb{R}^3$. This is the STANDARD BASIS for $\mathbb{R}^3$. In elementary calculus texts these vectors are usually called **i**, **j**, and **k**, respectively.

More generally, in $\mathbb{R}^n$ for $1 \leq k \leq n$ let e^k be the n-tuple that is zero in every coordinate except the kth coordinate where is value is 1. Then $\{e^1, e^2, \ldots, e^n\}$ is the STANDARD BASIS for $\mathbb{R}^n$.

Example 1.7.3. The space $\mathcal{P}_n(J)$ of polynomial functions of degree strictly less than $n \in \mathbb{N}$ on some interval $J \subseteq \mathbb{R}$ with nonempty interior is a vector space of dimension n. For each positive integer $n = 0, 1, 2, \ldots$ let $p_n(t) = t^n$ for all $t \in J$. Then $\{p_0, p_1, p_2, \ldots, p_{n-1}\}$ is a basis for $\mathcal{P}_n(J)$. We take this to be the STANDARD BASIS for $\mathcal{P}_n(J)$.

Example 1.7.4. The space $\mathcal{P}(J)$ of polynomial functions on some interval $J \subseteq \mathbb{R}$ with nonempty interior is an infinite dimensional vector space. For each $n = 0, 1, 2, \ldots$ define a function p_n on J by $p_n(x) = x^n$. Then the set $\{p_0, p_1, p_2, \ldots\}$ is a basis for the vector space $\mathcal{P}(J)$ of polynomial functions on J. We take this to be the STANDARD BASIS for $\mathcal{P}(J)$.

Example 1.7.5. Let $\mathfrak{M}_{m \times n}$ be the vector space of all $m \times n$ matrices of real numbers. For $1 \leq i \leq m$ and $1 \leq j \leq n$ let E^{ij} be the $m \times n$ matrix whose entry in the ith row and j^{th} column is 1 and all of whose other entries are 0. Then $\{E^{ij} : 1 \leq i \leq m \text{ and } 1 \leq j \leq n\}$ is a basis for $\mathfrak{M}_{m \times n}$.

Exercise 1.7.6. A 2×2 matrix $\begin{bmatrix} a & b \\ c & d \end{bmatrix}$ has *zero trace* if $a + d = 0$. Show that the set of all such matrices is a subspace of $\mathfrak{M}_{2 \times 2}$ and find a basis for it.

Exercise 1.7.7. Let $\mathcal{U}$ be the set of all matrices of real numbers of the form $\begin{bmatrix} u & -u - x \\ 0 & x \end{bmatrix}$ and $\mathcal{V}$ be the set of all real matrices of the form $\begin{bmatrix} v & 0 \\ w & -v \end{bmatrix}$. Find bases for $\mathcal{U}$, $\mathcal{V}$, $\mathcal{U} + \mathcal{V}$, and $\mathcal{U} \cap \mathcal{V}$.

To show that every nontrivial vector space has a basis we need to invoke *Zorn's lemma*, a set theoretic assumption that is equivalent to the *axiom of choice*. To this end we need to know about such things as partial orderings, chains, and maximal elements.

Definition 1.7.8. A RELATION on a set S is a subset of the Cartesian product $S \times S$. If the relation is denoted by $\leq$, then it is conventional to write $x \leq y$ (or equivalently, $y \geq x$) rather than $(x, y) \in \leq$.

Definition 1.7.9. A relation $\leq$ on a set S is REFLEXIVE if $x \leq x$ for all $x \in S$. It is TRANSITIVE if $x \leq z$ whenever $x \leq y$ and $y \leq z$. It is ANTISYMMETRIC if $x = y$ whenever $x \leq y$ and $y \leq x$. A relation that is reflexive, transitive, and antisymmetric is a PARTIAL ORDERING. A PARTIALLY ORDERED SET is a set on which a partial ordering has been defined.

Example 1.7.10. The set $\mathbb{R}$ of real numbers is a partially ordered set under the usual relation $\leq$.

Example 1.7.11. A family $\mathfrak{A}$ of subsets of a set S is a partially ordered set under the relation $\subseteq$. When $\mathfrak{A}$ is ordered in this fashion it is said to be ORDERED BY INCLUSION.

Example 1.7.12. Let $\mathcal{F}(S)$ be the family of real valued functions defined on a set S. For $f, g \in \mathcal{F}(S)$ write $f \leq g$ if $f(x) \leq g(x)$ for every $x \in S$. This is a partial ordering on $\mathcal{F}(S)$. It is known as POINTWISE ORDERING.

Definition 1.7.13. Let A be a subset of a partially ordered set S. An element $u \in S$ is an UPPER BOUND for A if $a \leq u$ for every $a \in A$. An element m in the partially ordered set S is MAXIMAL if there is no element of the set that is strictly greater than m; that is, m is maximal if $c = m$ whenever $c \in S$ and $c \geq m$. An element m in S is the LARGEST element of S if $m \geq s$ for every $s \in S$.

Similarly an element $l \in S$ is a LOWER BOUND for A if $l \leq a$ for every $a \in A$. An element m in the partially ordered set S is MINIMAL if there is no element of the set that is strictly less than m; that is, m is minimal if $c = m$ whenever $c \in S$ and $c \leq m$. An element m in S is the SMALLEST element of S if $m \leq s$ for every $s \in S$.

Example 1.7.14. Let $S = \{a, b, c\}$ be a three-element set. The family $\mathfrak{P}(S)$ of all subsets of S is partially ordered by inclusion. Then S is the largest element of $\mathfrak{P}(S)$ — and, of course, it is also a maximal element of $\mathfrak{P}(S)$. The family $\mathfrak{Q}(S)$ of all proper subsets of S has no largest element; but it has three maximal elements $\{b, c\}$, $\{a, c\}$, and $\{a, b\}$.

Proposition 1.7.15. *A linearly independent subset of a vector space V is a basis for V if and only if it is a maximal linearly independent subset.*

Proposition 1.7.16. *A spanning subset for a nontrivial vector space V is a basis for V if and only if it is a minimal spanning set for V.*

Definition 1.7.17. Let S be a partially ordered set with partial ordering $\leq$.

(1) Elements x and y in S are COMPARABLE if either $x \leq y$ or $y \leq x$.
(2) If $\leq$ is a partial ordering with respect to which any two elements of S are comparable, it is a LINEAR ORDERING (or a TOTAL ORDERING) and S is a LINEARLY ORDERED SET.
(3) A linearly ordered subset of S is a CHAIN in S.

Axiom 1.7.18 (Zorn's lemma). *A partially ordered set in which every chain has an upper bound has a maximal element.*

Theorem 1.7.19. *Let A be a linearly independent subset of a vector space V. Then there exists a basis for V that contains A.*

Hint for proof. Order the set of linearly independent subsets of V that contain A by inclusion. Apply *Zorn's lemma.*

Corollary 1.7.20. *Every vector space has a basis.*

Note that the empty set is a basis for the trivial vector space. (See Remark 1.6.6. Had we chosen the "constructive" definition for *span* we would have to say "nontrivial vector space" in the preceding corollary.)

Proposition 1.7.21. *Let B be a basis for a vector space V. Every element of V can be written in a unique way as a linear combination of elements of B.*

Notation 1.7.22. Let B be a basis for a vector space V over a field $\mathbb{F}$ and $x \in V$. By the preceding proposition there exists a unique finite set S of vectors in B and for each element e in S there exists a unique scalar x_e such that $x = \sum_{e \in S} x_e e$. If we agree to let $x_e = 0$ whenever $e \in B \setminus S$, we can just as well write $x = \sum_{e \in B} x_e e$. Although this notation may make it appear as if we are summing over an arbitrary, perhaps uncountable, set, the fact of the matter is that all but finitely many of the terms are zero. The function $x \colon B \to \mathbb{F} \colon e \mapsto x_e$ has finite support, so no "convergence" problems arise. Treat $\sum_{e \in B} x_e e$ as a finite sum. Associativity and commutativity of addition in V make the expression unambiguous.

Notice in the preceding that the symbol "x" ends up denoting two different things: a vector in V and a function in $l_c(B, \mathbb{F})$. We show in Proposition 2.2.10 that this identification is harmless. It is a good idea to teach

yourself to feel comfortable with identifying these two objects whenever you
are dealing with a vector space *with a basis.*

Notation 1.7.23. In finite dimensional vector spaces it is usual to adopt
some special notational conventions. Let V be an n-dimensional vector
space with an ordered basis $\{e^1, e^2, \ldots, e^n\}$. If $x \in V$, then by Proposition 1.7.21 we know that there are unique scalars $x_{e^1}, x_{e^2}, \ldots, x_{e^n}$ such
that

$$x = \sum_{k=1}^{n} x_{e^k} e^k.$$

The notation can be unambiguously simplified by writing

$$x = \sum_{k=1}^{n} x_k \, e^k.$$

Since the scalars $x_1, x_2, \ldots, x_n$ uniquely determine the vector x it has
become standard to write

$$x = (x_1, x_2, \ldots, x_n) \qquad \text{or} \qquad x = \begin{bmatrix} x_1 \\ x_2 \\ \vdots \\ x_n \end{bmatrix}.$$

That is, a vector x in a finite dimensional space with an ordered basis may
be represented as an n-tuple or as an $n \times 1$-matrix. The first of these is
frequently referred to as a *row vector* and the second as a *column vector.*

Next we verify that every subspace has a complementary subspace.

Proposition 1.7.24. *Let M be a subspace of a vector space V. Then there
exists a subspace N of V such that $V = M \oplus N$.*

Lemma 1.7.25. *Let V be a vector space with a finite basis $\{e^1, \ldots, e^n\}$
and let $v = \sum_{k=1}^{n} \alpha_k e^k$ be a vector in V. If $p \in \mathbb{N}_n$ and $\alpha_p \neq 0$, then
$\{e^1, \ldots, e^{p-1}, v, e^{p+1}, \ldots, e^n\}$ is a basis for V.*

Proposition 1.7.26. *If some basis for a vector space V contains n
elements, then every linearly independent subset of V with n elements is
also a basis.*

Hint for proof. Suppose $\{e^1, \ldots, e^n\}$ is a basis for V and $\{v^1, \ldots, v^n\}$ is linearly independent in V. Start by using Lemma 1.7.25 to show that (after perhaps renumbering the e^ks) the set $\{v^1, e^2, \ldots, e^n\}$ is a basis for V.

Corollary 1.7.27. *If a vector space V has a finite basis B, then every basis for V is finite and contains the same number of elements as B.*

Definition 1.7.28. A vector space is FINITE DIMENSIONAL if it has a finite basis and the DIMENSION of the space is the number of elements in this (hence any) basis for the space. The dimension of a finite dimensional vector space V is denoted by $\dim V$. If the space does not have a finite basis, it is INFINITE DIMENSIONAL.

Corollary 1.7.27 can be generalized to arbitrary vector spaces.

Theorem 1.7.29. *If B and C are bases for a vector space V, then B and C are cardinally equivalent; that is, there exists a bijection from B onto C.*

Proof. See [29], page 45, Theorem 1.12.   □

Proposition 1.7.30. *Let V be a vector space and suppose that $V = U \oplus W$. Prove that if B is a basis for U and C is a basis for W, then $B \cup C$ is a basis for V. From this conclude that $\dim V = \dim U + \dim W$.*

Definition 1.7.31. The TRANSPOSE of an $n \times n$ matrix $A = \begin{bmatrix} a_{ij} \end{bmatrix}$ is the matrix $A^t = \begin{bmatrix} a_{ji} \end{bmatrix}$ obtained by interchanging the rows and columns of A. The matrix A is SYMMETRIC if $A^t = A$.

Exercise 1.7.32. Let $\mathcal{S}_3$ be the vector space of all symmetric 3×3 matrices of real numbers.

(a) What is the dimension of $\mathcal{S}_3$?
(b) Find a basis for $\mathcal{S}_3$.

Chapter 2

LINEAR TRANSFORMATIONS

2.1. Linearity

Definition 2.1.1. Let V and W be vector spaces over the same field $\mathbb{F}$. A function $T\colon V \to W$ is LINEAR if $T(x + y) = Tx + Ty$ and $T(\alpha x) = \alpha Tx$ for all $x, y \in V$ and $\alpha \in \mathbb{F}$. For linear functions it is a matter of convention to write Tx instead of $T(x)$ whenever it does not cause confusion. (Of course, we would not write $Tx + y$ when we intend $T(x + y)$.) Linear functions are frequently called *linear transformations* or *linear maps*.

Notation 2.1.2. If V and W are vector spaces (over the same field $\mathbb{F}$) the family of all linear functions from V into W is denoted by $\mathfrak{L}(V, W)$. Linear functions are frequently called *linear transformations*, *linear maps*, or *linear mappings*. When $V = W$ we condense the notation $\mathfrak{L}(V, V)$ to $\mathfrak{L}(V)$ and we call the members of $\mathfrak{L}(V)$ *operators*.

Example 2.1.3. Let V and W be vector spaces over a field $\mathbb{F}$. For S, $T \in \mathfrak{L}(V, W)$ define $S + T$ by

$$(S + T)(x) := Sx + Tx$$

for all $x \in V$. For $T \in \mathfrak{L}(V, W)$ and $\alpha \in \mathbb{F}$ define αT by

$$(\alpha T)(x) := \alpha(Tx)$$

for all $x \in V$. Under these operations $\mathfrak{L}(V, W)$ is a vector space.

Proposition 2.1.4. *If $S\colon V \to W$ and $T\colon W \to X$ are linear maps between vector spaces, then the composite of these two functions, $T \circ S$, is a linear map from V into X.*

Convention 2.1.5. If $S\colon V \to W$ and $T\colon W \to X$ are linear maps between vector spaces, then the composite linear map of these two functions is nearly always written as TS rather than $T \circ S$. In the same vein, $T^2 = T \circ T$, $T^3 = T \circ T \circ T$, and so on.

Exercise 2.1.6. Use the notation of Definition 1.4.1 and suppose that $(V, +, M)$ and $(W, +, M)$ are vector spaces over a common field $\mathbb{F}$ and that $T \in \mathrm{Hom}(V, W)$ is such that the diagram

$$
\begin{array}{ccc}
V & \xrightarrow{\;T\;} & W \\
\downarrow{\scriptstyle M_\alpha} & & \downarrow{\scriptstyle M_\alpha} \\
V & \xrightarrow[\;T\;]{} & W
\end{array}
$$

commutes for every $\alpha \in \mathbb{F}$. What can be said about the homomorphism T?

Example 2.1.7. Let $a < b$ and $\mathcal{C} = \mathcal{C}([a,b])$ be the vector space of all continuous real valued functions on the interval $[a,b]$. Then integration

$$
T\colon \mathcal{C} \to \mathbb{R}\colon f \mapsto \int_a^b f(t)\,dt
$$

is a linear map.

Example 2.1.8. Let $a < b$ and $\mathcal{C}^1 = \mathcal{C}^1([a,b])$ be the set of all continuously differentiable real valued functions on the interval $[a,b]$. (Recall that a function is CONTINUOUSLY DIFFERENTIABLE if it has a derivative and the derivative is continuous.) Then differentiation

$$
D\colon \mathcal{C}^1 \to \mathcal{C}\colon f \mapsto f'
$$

is linear.

Example 2.1.9. Let $\mathbb{R}^\infty$ be the vector space of all sequences of real numbers and define

$$
S\colon \mathbb{R}^\infty \to \mathbb{R}^\infty\colon (x_1, x_2, x_3, \dots) \mapsto (0, x_1, x_2, \dots).
$$

Then S is a linear operator. It is called the UNILATERAL SHIFT OPERATOR.

Definition 2.1.10. Let $T\colon V \to W$ be a linear transformation between vector spaces. Then $\ker T$, the KERNEL (or NULLSPACE) of T is defined to be the set of all x in V such that $Tx = \mathbf{0}$. Also, $\operatorname{ran} T$, the RANGE of T (or the IMAGE of T), is the set of all y in W such that $y = Tx$ for some x in V. The RANK of T is the dimension of its range and the NULLITY of T is the dimension of its kernel.

Definition 2.1.11. Let $T: V \to W$ be a linear transformation between vector spaces and let A be a subset of V. Define $T^{\to}(A) := \{Tx : x \in A\}$. This is the (DIRECT) IMAGE OF A UNDER T.

Proposition 2.1.12. *Let $T: V \to W$ be a linear map between vector spaces and $M \preceq V$. Then $T^{\to}(M)$ is a subspace of W. In particular, the range of a linear map is a subspace of the codomain of the map.*

Definition 2.1.13. Let $T: V \to W$ be a linear transformation between vector spaces and let B be a subset of W. Define $T^{\leftarrow}(B) := \{x \in V : Tx \in B\}$. This is the INVERSE IMAGE OF B UNDER T.

Proposition 2.1.14. *Let $T: V \to W$ be a linear map between vector spaces and $M \preceq W$. Then $T^{\leftarrow}(M)$ is a subspace of V. In particular, the kernel of a linear map is a subspace of the domain of the map.*

Exercise 2.1.15. Let $T: \mathbb{R}^3 \to \mathbb{R}^3 : x = (x_1, x_2, x_3) \mapsto (x_1 + 3x_2 - 2x_3, x_1 - 4x_3, x_1 + 6x_2)$.

(a) Identify the kernel of T by describing it geometrically and by giving its equation(s).
(b) Identify the range of T by describing it geometrically and by giving its equation(s).

Exercise 2.1.16. Let T be the linear map from $\mathbb{R}^3$ to $\mathbb{R}^3$ defined by

$$T(x, y, z) = (2x + 6y - 4z, 3x + 9y - 6z, 4x + 12y - 8z).$$

Describe the kernel of T geometrically and give its equation(s). Describe the range of T geometrically and give its equation(s).

Exercise 2.1.17. Let $\mathcal{C} = \mathcal{C}[a, b]$ be the vector space of all continuous real valued functions on the interval $[a, b]$ and $\mathcal{C}^1 = \mathcal{C}^1[a, b]$ be the vector space of all continuously differentiable real valued functions on $[a, b]$. Let $D: \mathcal{C}^1 \to \mathcal{C}$

be the linear transformation defined by

$$Df = f'$$

and let $T: \mathcal{C} \to \mathcal{C}^1$ be the linear transformation defined by

$$(Tf)(x) = \int_a^x f(t)\,dt$$

for all $f \in \mathcal{C}$ and $x \in [a, b]$.

(a) Compute (and simplify) $(DTf)(x)$.
(b) Compute (and simplify) $(TDf)(x)$.
(c) Find the kernel of T.
(d) Find the range of T.

Proposition 2.1.18. *A linear map $T: V \to W$ between vector spaces is injective if and only if $\ker T = \{0\}$.*

2.2. Invertible Linear Maps

Notation 2.2.1. In the sequel we will usually denote the identity operator $x \mapsto x$ on a vector space V by I_V, or just I, rather than by id_V.

Definition 2.2.2. A linear map $T: V \to W$ between vector spaces is LEFT INVERTIBLE (or has a LEFT INVERSE, or is a SECTION) if there exists a linear map $L: W \to V$ such that $LT = I_V$.

The map T is RIGHT INVERTIBLE (or has a RIGHT INVERSE, or is a RETRACTION) if there exists a linear map $R: W \to V$ such that $TR = I_W$. We say that T is INVERTIBLE (or has an INVERSE, or is an ISOMORPHISM) if there exists a linear map $T^{-1}: W \to V$ which is both a left and a right inverse for T. If there exists an isomorphism between two vector spaces V and W, we say that the spaces are ISOMORPHIC and we write $V \cong W$.

Exercise 2.2.3. Show that an operator $T \in \mathfrak{L}(V)$ is invertible if it satisfies the equation

$$T^2 - T + I_V = \mathbf{0}.$$

Example 2.2.4. The unilateral shift operator S on the vector space $\mathbb{R}^\infty$ of all sequences of real numbers (see 2.1.9) is injective but not surjective. It is left invertible but not right invertible.

Proposition 2.2.5. *A linear map $T\colon V \to W$ between vector spaces is invertible if and only if has both a left inverse and a right inverse.*

Proposition 2.2.6. *A linear map between vector spaces is invertible if and only if it is bijective.*

Proposition 2.2.7. *Let $T\colon V \to W$ and $S\colon W \to X$ be linear maps between vector spaces. If T and S are invertible, then so is ST and $(ST)^{-1} = T^{-1}S^{-1}$.*

Proposition 2.2.8. *An operator T on a vector space V is invertible if and only if it has a unique right inverse.*

Hint for proof. Consider $ST + S - I_V$, where S is the unique right inverse for T.

Example 2.2.9. Every n-dimensional real vector space is isomorphic to $\mathbb{R}^n$.

Hint for proof. Recall the notational conventions made in 1.7.23.

Example 2.2.10. Let B be a basis for a vector space V over a field $\mathbb{F}$. Then $V \cong l_c(B, \mathbb{F})$.

Hint for proof. Recall the notational conventions made in 1.7.22.

Proposition 2.2.11. *Let $S, T \in \mathfrak{L}(V, W)$ where V and W are vector spaces over a field $\mathbb{F}$; and let B be a basis for V. If $S(e) = T(e)$ for every $e \in B$, then $S = T$.*

Proposition 2.2.12. *Let V and W be a vector spaces. If $V = M \oplus N$ and $T\colon M \to W$ is a linear map, then there exists $\widehat{T} \in \mathfrak{L}(V, W)$ such that $\widehat{T}\big|_M = T$ and $\widehat{T}\big|_N = \mathbf{0}$.*

We can now make a considerable improvement on Proposition 2.2.6.

Proposition 2.2.13. *A linear transformation has a left inverse if and only if it is injective.*

Proposition 2.2.14. *A linear transformation has a right inverse if and only if it is surjective.*

Proposition 2.2.15. *Let V and W be vector spaces over a field $\mathbb{F}$ and B be a basis for V. If $f\colon B \to W$, then there exists a unique linear map $T_f\colon V \to W$ that is an extension of f.*

Exercise 2.2.16. Let S be a set, V be a vector space over a field $\mathbb{F}$, and $f: S \to V$ be a bijection. Explain how to use f to make S into a vector space isomorphic to V.

2.3. Matrix Representations

Proposition 2.3.1. *Let* $T \in \mathcal{L}(V, W)$ *where* V *is an* n-*dimensional vector space and* W *is an* m-*dimensional vector space and let* $\{e^1, e^2, \ldots, e^n\}$ *be an ordered basis for* V. *Define an* $m \times n$-*matrix* $[T]$ *whose* k^{th} *column (1 $\leq$ $k \leq n$) is the column vector* Te^k *(see 1.7.23). Then for each* $x \in V$ *we have*

$$Tx = [T]x.$$

The displayed equation above requires a little interpretation. The left side is T evaluated at x; the right side is an $m \times n$ matrix multiplied by an $n \times 1$ matrix (that is, a column vector). Then the asserted equality is of two $m \times 1$ matrices (column vectors).

Definition 2.3.2. If V and W are finite dimensional vector spaces with ordered bases and $T \in \mathcal{L}(V, W)$, then the matrix $[T]$ in the preceding proposition is the MATRIX REPRESENTATION of T.

Exercise 2.3.3. Let $T : \mathbb{R}^4 \to \mathbb{R}^3$ be defined by

$$Tx = (x_1 - 3x_3 + x_4, 2x_1 + x_2 + x_3 + x_4, 3x_2 - 4x_3 + 7x_4)$$

for every $x = (x_1, x_2, x_3, x_4) \in \mathbb{R}^4$.

(a) Find $[T]$.
(b) Find $T(1, -2, 1, 3)$.
(c) Independently of part (b) calculate the matrix product $[T] \begin{bmatrix} 1 \\ -2 \\ 1 \\ 3 \end{bmatrix}$.
(d) Find $\ker T$.
(e) Find $\operatorname{ran} T$.

Exercise 2.3.4. Let $\mathcal{P}_4(\mathbb{R})$ be the vector space of polynomial functions of degree strictly less than 4 on $\mathbb{R}$. Consider the linear transformation $D^2 : \mathcal{P}_4 \to \mathcal{P}_4 : f \mapsto f''$.

(a) Find the matrix representation of D^2 (with respect to the standard basis for $\mathcal{P}^4(\mathbb{R})$).
(b) Find the kernel of D^2.
(c) Find the range of D^2.

Exercise 2.3.5. Let $T\colon \mathcal{P}_4(\mathbb{R}) \to \mathcal{P}_5(\mathbb{R})$ be the linear transformation defined by $(Tp)(t) = (2 + 3t)p(t)$ for every $p \in \mathcal{P}_4(\mathbb{R})$ and $t \in \mathbb{R}$. Find the matrix representation of T with respect to the standard bases for $\mathcal{P}_4(\mathbb{R})$ and $\mathcal{P}_5(\mathbb{R})$.

Exercise 2.3.6. Let $\mathcal{P}_n(\mathbb{R})$ be the vector space of all polynomial functions on $\mathbb{R}$ with degree strictly less than n. Define $T\colon \mathcal{P}_3(\mathbb{R}) \to \mathcal{P}_5(\mathbb{R})$ by

$$Tf(x) = \int_0^x \int_0^u p(t)\, dt\, du$$

for all $x, u \in \mathbb{R}$.

(a) Find the matrix representation of the linear map T (with respect to the standard bases for $\mathcal{P}_3(\mathbb{R})$ and $\mathcal{P}_5(\mathbb{R})$).
(b) Find the kernel of T.
(c) Find the range of T.

2.4. Spans, Independence, and Linearity

Proposition 2.4.1. *Let $T\colon V \to W$ be an injective linear map between vector spaces. If A is a subset of V such that $T^{\to}(A)$ is linearly independent, then A is linearly independent.*

Proposition 2.4.2. *Let $T\colon V \to W$ be an injective linear map between vector spaces. If A is a linearly independent subset of V, then $T^{\to}(A)$ is a linearly independent subset of W.*

Proposition 2.4.3. *Let $T\colon V \to W$ be a linear map between vector spaces and $A \subseteq V$. Then $T^{\to}(\operatorname{span} A) = \operatorname{span} T^{\to}(A)$.*

Proposition 2.4.4. *Let $T\colon V \to W$ be an injective linear map between vector spaces. If B is a basis for a subspace U of V, then $T^{\to}(B)$ is a basis for $T^{\to}(U)$.*

Proposition 2.4.5. *Let $T\colon V \to W$ be an injective linear map between vector spaces. If V is spanned by a set B of vectors and $T^{\to}(B)$ is a basis for W, then B is a basis for V and T is an isomorphism.*

Exercise 2.4.6. Prove that a linear transformation $T\colon \mathbb{R}^3 \to \mathbb{R}^2$ cannot be one-to-one and that a linear transformation $S\colon \mathbb{R}^2 \to \mathbb{R}^3$ cannot be onto. What is the most general version of these assertions that you can invent (and prove)?

Proposition 2.4.7. *Suppose that V and W are finite dimensional vector spaces of the same finite dimension and that $T\colon V \to W$ is a linear map. Then the following are equivalent;*

(a) *T is injective;*
(b) *T is surjective; and*
(c) *T is invertible.*

2.5. Dual Spaces

Definition 2.5.1. Let V be a vector space over a field $\mathbb{F}$. A linear map $f\colon V \to \mathbb{F}$ is a LINEAR FUNCTIONAL on V. The set of linear functionals on V is denoted by V^*; that is, $V^* = \mathfrak{L}(V, \mathbb{F})$. The vector space V^* is the DUAL SPACE of V.

Convention 2.5.2. Let B be a basis for a vector space V over a field $\mathbb{F}$. Recall that in 1.7.22 we adopted the notation

$$x = \sum_{e \in B} x_e e$$

where x denotes both an element of V and a scalar valued function on B with finite support. In Example 2.2.10 we justified this identification by establishing that the vector spaces V and $l_c(B, \mathbb{F})$ are isomorphic. Notice that this is an extension of the usual notation in $\mathbb{R}^n$ where we write a vector $\mathbf{v}$ in terms of its components:

$$x = \sum_{k=1}^{n} x_k\, e^k.$$

Exercise 2.5.3. According to Convention 2.5.2 above, what is the value of f_e when e and f are elements of the basis B?

Proposition 2.5.4. *Let V be a vector space with basis B. For every $v \in V$ define a function v^* on V by*

$$v^*(x) = \sum_{e \in B} x_e\, v_e \qquad \text{for all } x \in V.$$

Then v^* is a linear functional on V.

Notation 2.5.5. In the preceding Proposition 2.5.4 the value $v^*(x)$ of v^* at x is often written as $\langle x, v \rangle$.

Exercise 2.5.6. Consider the notation 2.5.5 above in the special case that the scalar field $\mathbb{F} = \mathbb{R}$. Then $\langle\ ,\ \rangle$ is an inner product on the vector space V. (For a definition of *inner product* see 5.1.1.)

Exercise 2.5.7. In the special case that the scalar field $\mathbb{F} = \mathbb{C}$, things above are usually done a bit differently. For $v \in V$ the function v^* is defined by

$$v^*(x) = \langle x, v \rangle = \sum_{e \in B} x_e\, \overline{v_e}\,.$$

Why do you think things are done this way?

Proposition 2.5.8. *Let v be a nonzero vector in a vector space V and B be a basis for V that contains the vector v. Then there exists a linear functional $f \in V^*$ such that $f(v) = 1$ and $f(e) = 0$ for every $e \in B \setminus \{v\}$.*

Corollary 2.5.9. *Let M be a subspace of a vector space V and v a vector in V that does not belong to M. Then there exists $f \in V^*$ such that $f(v) = 1$ and $f^{\rightarrow}(M) = \{0\}$.*

Corollary 2.5.10. *If v is a vector in a vector space V and $f(v) = 0$ for every $f \in V^*$, then $v = \mathbf{0}$.*

Definition 2.5.11. Let $\mathbb{F}$ be a field. A family $\mathcal{F}$ of $\mathbb{F}$-valued functions on a set S containing at least two points SEPARATES POINTS OF S if for every $x, y \in S$ such that $x \neq y$ there exists $f \in \mathcal{F}$ such that $f(x) \neq f(y)$.

Corollary 2.5.12. *For every nontrivial vector space V, the dual space V^* separates points of V.*

Proposition 2.5.13. *Let V be a vector space with basis B. The map $\Phi\colon V \to V^*\colon v \mapsto v^*$ (see Proposition 2.5.4) is linear and injective.*

The next result is the *Riesz-Fréchet theorem for finite dimensional vector spaces with basis*. It is important to keep in mind that the result does *not* hold for infinite dimensional vector spaces (see Proposition 2.5.18) and that the mapping Φ depends on the basis which has been chosen for the vector space.

Theorem 2.5.14. *Let V be a finite dimensional vector space with basis B. Then the map Φ defined in the preceding Proposition 2.5.13 is an isomorphism. Thus for every $f \in V^*$ there exists a unique vector $a \in V$ such that $a^* = f$.*

Definition 2.5.15. Let V be a vector space with basis $\{e^\lambda\colon \lambda \in \Lambda\}$. A basis $\{\varepsilon^\lambda\colon \lambda \in \Lambda\}$ for V^* is the DUAL BASIS for V^* if it satisfies

$$\varepsilon^\mu(e^\lambda) = \begin{cases} 1, & \text{if } \mu = \lambda; \\ 0, & \text{if } \mu \neq \lambda. \end{cases}$$

Theorem 2.5.16. *Every finite dimensional vector space V with a basis has a unique dual basis for its dual space. In fact, if $\{e^1, \ldots, e^n\}$ is a basis for V, then $\{(e^1)^*, \ldots, (e^n)^*\}$ is the dual basis for V^*.*

Corollary 2.5.17. *If a vector space V is finite dimensional, then so is its dual space and $\dim V = \dim V^*$.*

In Proposition 2.5.13 we showed that the map

$$\Phi\colon V \to V^*\colon v \mapsto v^*$$

is always an injective linear map. In Corollary 2.5.17 we showed that if V is finite dimensional, then so is V^* and Φ is an isomorphism between V and V^*. This is never true in infinite dimensional spaces.

Proposition 2.5.18. *If V is infinite dimensional, then Φ is not an isomorphism.*

Hint for proof. Let B be a basis for V. Is there a functional $g \in V^*$ such that $g(e) = 1$ for every $e \in B$? Could such a functional be $\Phi(x)$ for some $x \in V$?

Proposition 2.5.19. *Let V be a vector space over a field $\mathbb{F}$. For every x in V define*

$$\widehat{x}\colon V^* \to \mathbb{F}\colon \phi \mapsto \phi(x).$$

(a) *The vector $\widehat{x}$ belongs to V^{**} for each $x \in V$.*
(b) *Let Γ_V be the map from V to V^{**} that takes x to $\widehat{x}$. (When no confusion is likely we write Γ for Γ_V, so that $\Gamma(x) = \widehat{x}$ for each $x \in V$.) The function Γ is linear.*
(c) *The function Γ is injective.*

Proposition 2.5.20. *If V is a finite dimensional vector space, then the map $\Gamma\colon V \to V^{**}$ defined in the preceding Proposition 2.5.19 is an isomorphism.*

Proposition 2.5.21. *If V is infinite dimensional, then the mapping Γ (defined in 2.5.19) is not an isomorphism.*

Hint for proof. Let B be a basis for V and $\psi \in V^*$ be as in Proposition 2.5.18. Show that if we let C_0 be $\{e^* : e \in B\}$, then the set $C_0 \cup \{\psi\}$ is linearly independent and can therefore be extended to a basis C for V^*. Find an element τ in V^{**} such that $\tau(\psi) = 1$ and $\tau(\phi) = 0$ for every other $\phi \in C$. Can τ be Γx for some $x \in V$?

2.6. Annihilators

Notation 2.6.1. Let V be a vector space and $M \subseteq V$. Then

$$M^\perp := \{f \in V^* : f(x) = 0 \text{ for all } x \in M\}$$

We say that $M^\perp$ is the ANNIHILATOR of M. (The reasons for using the familiar "orthogonal complement" notation $M^\perp$ (usually read as "M perp") will become apparent when we study inner product spaces, where "orthogonality" actually makes sense.)

Exercise 2.6.2. Find the annihilator in $(\mathbb{R}^2)^*$ of the vector $(1, 1)$ in $\mathbb{R}^2$. (Express your answer in terms of the standard dual basis for $(\mathbb{R}^2)^*$.)

Proposition 2.6.3. *Let M and N be subsets of a vector space V. Then*

(a) *$M^\perp$ is a subspace of V^*.*
(b) *If $M \subseteq N$, then $N^\perp \preceq M^\perp$.*
(c) *$(\text{span } M)^\perp = M^\perp$.*
(d) *$(M \cup N)^\perp = M^\perp \cap N^\perp$.*

Proposition 2.6.4. *If M and N be subspaces of a vector space V, then*

$$(M + N)^\perp = M^\perp \cap N^\perp.$$

Exercise 2.6.5. Explain why it is necessary in the preceding proposition to assume that M and N are subspaces of V and not just subsets of V.

Notation 2.6.6. Let V be a vector space and $F \subseteq V^*$. Then

$$F_\perp := \{x \in V : f(x) = 0 \text{ for all } f \in F\}$$

We say that $F_\perp$ is the PRE-ANNIHILATOR of F.

Proposition 2.6.7. *If M is a subspace of a vector space V, then $(M^\perp)_\perp = M$.*

Exercise 2.6.8. Propositions 2.6.3 and 2.6.4 asserted some properties of the annihilator mapping $M \mapsto M^{\perp}$. See to what extent you can prove similar results about the pre-annihilator mapping $F \mapsto F_{\perp}$. What can you say about the set $\left(F_{\perp}\right)^{\perp}$?

Proposition 2.6.9. *Let V be a finite dimensional vector space and F be a subspace of V^*. If $F_{\perp} = \{0\}$, then $F = V^*$.*

Chapter 3

THE LANGUAGE OF
CATEGORIES

3.1. Objects and Morphisms

Definition 3.1.1. Let $\mathfrak{A}$ be a class, whose members we call OBJECTS. For every pair (S, T) of objects we associate a set $\mathfrak{Mor}(S, T)$, whose members we call MORPHISMS from S to T. We assume that $\mathfrak{Mor}(S, T)$ and $\mathfrak{Mor}(U, V)$ are disjoint unless $S = U$ and $T = V$.

We suppose further that there is an operation $\circ$ (called COMPOSITION) that associates with every $\alpha \in \mathfrak{Mor}(S, T)$ and every $\beta \in \mathfrak{Mor}(T, U)$ a morphism $\beta \circ \alpha \in \mathfrak{Mor}(S, U)$ in such a way that:

(1) $\gamma \circ (\beta \circ \alpha) = (\gamma \circ \beta) \circ \alpha$ whenever $\alpha \in \mathfrak{Mor}(S, T)$, $\beta \in \mathfrak{Mor}(T, U)$, and $\gamma \in \mathfrak{Mor}(U, V)$;
(2) for every object S there is a morphism $I_S \in \mathfrak{Mor}(S, S)$ satisfying $\alpha \circ I_S = \alpha$ whenever $\alpha \in \mathfrak{Mor}(S, T)$ and $I_S \circ \beta = \beta$ whenever $\beta \in \mathfrak{Mor}(R, S)$.

Under these circumstances the class $\mathfrak{A}$, together with the associated families of morphisms, is a CATEGORY.

We will reserve the notation $S \xrightarrow{\alpha} T$ for a situation in which S and T are objects in some category and α is a morphism belonging to $\mathfrak{Mor}(S, T)$. As is the case with groups and vector spaces we usually omit the composition symbol $\circ$ and write $\beta\alpha$ for $\beta \circ \alpha$.

Example 3.1.2. The category **SET** has sets for objects and functions (maps) as morphisms.

43

Example 3.1.3. The category **AbGp** has Abelian groups for objects and group homomorphisms as morphisms. (See Proposition 1.1.17.)

Example 3.1.4. The category **VEC** has vector spaces for objects and linear transformations as morphisms. (See Proposition 2.1.4.)

Example 3.1.5. Let S and T be partially ordered sets. A function $f \colon S \to T$ is ORDER PRESERVING if $f(x) \le f(y)$ in T whenever $x \le y$ in S. The category **POSET** has partially ordered sets for objects and order preserving maps as morphisms.

The preceding examples are examples of *concrete categories* — that is, categories in which the objects are sets (together, usually, with additional structure) and the morphism are functions (usually preserving, in some sense, this extra structure). In these notes the categories of interest are concrete ones. Even so, it may be of interest to see an example of a category that is *not* concrete.

Example 3.1.6. Let G be a monoid. Consider a category $\mathbf{C}_G$ having exactly one object, which we call $\star$. Since there is only one object there is only one family of morphisms $\mathfrak{Mor}(\star, \star)$, which we take to be G. Composition of morphisms is defined to be the monoid multiplication. That is, $a \circ b := ab$ for all $a,\ b \in G$. Clearly composition is associative and the identity element of G is the identity morphism. So $\mathbf{C}_G$ is a category.

Definition 3.1.7. In any concrete category we will call an injective morphism a MONOMORPHISM and a surjective morphism an EPIMORPHISM.

Caution 3.1.8. The definitions above reflect the original Bourbaki use of the term and are the ones most commonly adopted by mathematicians outside of category theory where "monomorphism" means "left cancellable" and "epimorphism" means "right cancellable". (Notice that the terms *injective* and *surjective* may not make sense when applied to morphisms in a category that is not concrete.)

A morphism $B \xrightarrow{g} C$ is LEFT CANCELLABLE if whenever morphisms $A \xrightarrow{f_1} B$ and $A \xrightarrow{f_2} B$ satisfy $gf_1 = gf_2$, then $f_1 = f_2$. Saunders Mac Lane suggested calling left cancellable morphisms MONIC morphisms. The distinction between monic morphisms and monomorphisms turns out to be slight. In these notes almost all of the morphisms we encounter are monic if and only if they are monomorphisms. As an easy exercise prove that any injective morphism in a (concrete) category is monic. The converse sometimes fails.

In the same vein Mac Lane suggested calling a *right cancellable* morphism (that is, a morphism $A \xrightarrow{f} B$ such that whenever morphisms $B \xrightarrow{g_1} C$ and $B \xrightarrow{g_2} C$ satisfy $g_1 f = g_2 f$, then $g_1 = g_2$) an EPIC morphism. Again it is an easy exercise to show that in a (concrete) category any epimorphism is epic. The converse, however, fails in some rather common categories.

Definition 3.1.9. The terminology for inverses of morphisms in categories is essentially the same as for functions. Let $S \xrightarrow{\alpha} T$ and $T \xrightarrow{\beta} S$ be morphisms in a category. If $\beta \circ \alpha = I_S$, then β is a LEFT INVERSE of α and, equivalently, α is a RIGHT INVERSE of β. We say that the morphism α is an ISOMORPHISM (or is INVERTIBLE) if there exists a morphism $T \xrightarrow{\beta} S$ which is both a left and a right inverse for α. Such a function is denoted by α^{-1} and is called the INVERSE of α.

Proposition 3.1.10. *If a morphism in some category has both a left and a right inverse, then it is invertible.*

In any concrete category one can inquire whether every bijective morphism (that is, every map which is both a monomorphism and an epimorphism) is an isomorphism. We saw in Proposition 2.2.6 that in the category **VEC** the answer is *yes*. In the next example the answer is *no*.

Example 3.1.11. In the category **POSET** of partially ordered sets and order preserving maps not every bijective morphism is an isomorphism.

Example 3.1.12. If in the category $\mathbf{C}_G$ of Example 3.1.6 the monoid G is a group, then every morphism in $\mathbf{C}_G$ is an isomorphism.

3.2. Functors

Definition 3.2.1. If **A** and **B** are categories a (COVARIANT) FUNCTOR F from **A** to **B** (written $\mathbf{A} \xrightarrow{F} \mathbf{B}$) is a pair of maps: an OBJECT MAP F that associates with each object S in **A** an object $F(S)$ in **B** and a MORPHISM MAP (also denoted by F) that associates with each morphism

$f \in \mathfrak{Mor}(S,T)$ in **A** a morphism $F(f) \in \mathfrak{Mor}(F(S), F(T))$ in **B**, in such a way that

(1) $F(g \circ f) = F(g) \circ F(f)$ whenever $g \circ f$ is defined in **A**; and

(2) $F(\text{id}_S) = \text{id}_{F(S)}$ for every object S in **A**.

The definition of a CONTRAVARIANT FUNCTOR $\mathbf{A} \xrightarrow{F} \mathbf{B}$ differs from the preceding definition only in that, first, the morphism map associates with each morphism $f \in \mathfrak{Mor}(S,T)$ in **A** a morphism $F(f) \in \mathfrak{Mor}(F(T), F(S))$ in **B** and, second, condition (1) above is replaced by

(1′) $F(g \circ f) = F(f) \circ F(g)$ whenever $g \circ f$ is defined in **A**.

Example 3.2.2. A FORGETFUL FUNCTOR is a functor that maps objects and morphisms from a category **C** to a category **C′** with less structure or fewer properties. For example, if V is a vector space, the functor F that "forgets" about the operation of scalar multiplication on vector spaces would map V into the category of Abelian groups. (The Abelian group $F(V)$ would have the same set of elements as the vector space V and the same operation of addition, but it would have no scalar multiplication.) A linear map $T : V \to W$ between vector spaces would be taken by the functor F to a group homomorphism $F(T)$ between the Abelian groups $F(V)$ and $F(W)$.

Forgetful functors can "forget" about properties as well. If G is an object in the category of Abelian groups, the functor that "forgets" about commutativity in Abelian groups would take G into the category of groups.

It was mentioned in the preceding section that all the categories that are of interest in these notes are concrete categories (ones in which the objects are sets with additional structure and the morphisms are maps that preserve, in some sense, this additional structure). We will have several occasions to use a special type of forgetful functor — one that forgets about all the structure of the objects except the underlying set and that forgets any structure preserving properties of the morphisms. If A is an object in some concrete category **C**, we denote by $|A|$ its underlying set. And if $A \xrightarrow{f} B$ is a morphism in **C** we denote by $|f|$ the map from $|A|$ to $|B|$ regarded simply as a function between sets. It is easy to see that $|\ \ |$, which takes objects in **C** to objects in **SET** (the category of sets and maps) and morphisms in **C** to morphisms in **SET**, is a covariant functor.

In the category **VEC** of vector spaces and linear maps, for example, $|\ \ |$ causes a vector space V to "forget" about both its addition and scalar multiplication ($|V|$ is just a set). And if $T\colon V \to W$ is a linear transformation, then $|T|\colon |V| \to |W|$ is just a map between sets — it has "forgotten" about preserving the operations.

Notation 3.2.3. Let $f\colon S \to T$ be a function between sets. Then we define $f^{\to}(A) = \{f(x)\colon x \in A\}$ and $f^{\leftarrow}(B) = \{x \in S\colon f(x) \in B\}$. We say that $f^{\to}(A)$ is the IMAGE OF A UNDER f and that $f^{\leftarrow}(B)$ is the PREIMAGE OF B UNDER f.

Definition 3.2.4. A partially ordered set is ORDER COMPLETE if every nonempty subset has a supremum (that is, a least upper bound) and an infimum (a greatest lower bound).

Definition 3.2.5. Let S be a set. Then the POWER SET of S, denoted by $\mathfrak{P}(S)$, is the family of all subsets of S.

Example 3.2.6 (The power set functors). Let S be a nonempty set.

(a) The power set $\mathfrak{P}(S)$ of S partially ordered by $\subseteq$ is order complete.
(b) The class of order complete partially ordered sets and order preserving maps is a category.
(c) For each function f between sets let $\mathfrak{P}(f) = f^{\to}$. Then $\mathfrak{P}$ is a covariant functor from the category of sets and functions to the category of order complete partially ordered sets and order preserving maps.
(d) For each function f between sets let $\mathfrak{P}(f) = f^{\leftarrow}$. Then $\mathfrak{P}$ is a contravariant functor from the category of sets and functions to the category of order complete partially ordered sets and order preserving maps.

Definition 3.2.7. Let $T\colon V \to W$ be a linear map between vector spaces. For every $g \in W^*$ let $T^*(g) = gT$. Notice that $T^*(g) \in V^*$. The map T^* from the vector space W^* into the vector space V^* is the (vector space) ADJOINT map of T.

Caution 3.2.8. In inner product spaces we will use the same notation T^* for a different map. If $T\colon V \to W$ is a linear map between inner product spaces, then the (inner product space) adjoint transformation T^* maps W to V (not W^* to V^*).

Example 3.2.9 (The vector space duality functor). Let $T \in \mathfrak{L}(V,W)$ where V and W are vector spaces over a field $\mathbb{F}$. Then the pair of maps $V \mapsto V^*$ and $T \mapsto T^*$ is a contravariant functor from the category of vector spaces and linear maps into itself. Show that (the morphism map of) this functor is linear. (That is, show that $(S+T)^* = S^* + T^*$ and $(\alpha T)^* = \alpha T^*$ for all S, $T \in \mathfrak{L}(V, W)$ and $\alpha \in \mathbb{F}$.)

There are several quite different results that in various texts are labeled as *the fundamental theorem of linear algebra*. Many of them seem to me not to be particularly "fundamental" because they apply only to finite dimensional inner product spaces or, what amounts to the same thing, matrices. I feel the following result deserves the name because it holds for arbitrary linear maps between arbitrary vector spaces.

Theorem 3.2.10 (Fundamental Theorem of Linear Algebra). *For every linear map $T: V \to W$ between vector spaces the following hold.*

(1) $\ker T^* = (\operatorname{ran} T)^{\perp}$;
(2) $\operatorname{ran} T^* = (\ker T)^{\perp}$;
(3) $\ker T = (\operatorname{ran} T^*)_{\perp}$; *and*
(4) $\operatorname{ran} T = (\ker T^*)_{\perp}$.

Hint for proof. Showing in (2) that $(\ker T)^{\perp} \subseteq \operatorname{ran} T^*$ takes some thought. Let $f \in (\ker T)^{\perp}$. We wish to find a $g \in W^*$ such that $f = T^* g$. By Propositions 2.1.14 and 1.7.24 there exists a subspace M of V such that $V = \ker T \oplus M$. Let $\iota: M \to V$ be the inclusion mapping of M into V and $T|_M$ be the restriction of T to M. Use Proposition 2.2.13 to show that $T|_M$ has a left inverse $S: W \to M$. Let $g := f \iota S$. Use Theorem 1.5.20.

Exercise 3.2.11. What is the relationship between a linear map T being injective and its adjoint T^* being surjective? between T being surjective and T^* being injective?

3.3. Universal Mapping Properties

Much of mathematics involves the construction of new objects from old ones — things such as products, coproducts, quotients, completions, compactifications, and unitizations. Often it is possible — and highly desirable — to characterize such a construction by means of a diagram that describes what the constructed object "does" rather than telling what it "is" or how it is constructed. Such a diagram is a UNIVERSAL MAPPING DIAGRAM and it describes the UNIVERSAL MAPPING PROPERTY of the object being constructed.

Here is a first example of such a property.

Definition 3.3.1. Let F be an object in a concrete category $\mathbf{C}$ and $\iota\colon S \to$ $|F|$ be an inclusion map whose domain is a nonempty set S. We say that the object F is FREE ON the set S (or that F is the FREE OBJECT GENERATED BY S) if for every object A in $\mathbf{C}$ and every map $f\colon S \to |A|$ there exists a unique morphism $\widetilde{f}_\iota\colon F \to A$ in $\mathbf{C}$ such that $|\widetilde{f}_\iota| \circ \iota = f$.

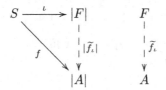

We will be interested in FREE VECTOR SPACES; that is, free objects in the category **VEC** of vector spaces and linear maps. Naturally, merely *defining* a concept does not guarantee its existence. It turns out, in fact, that free vector spaces exist on arbitrary sets. (See Exercise 3.3.5.)

Exercise 3.3.2. In the preceding definition reference to the forgetful functor is often omitted and the accompanying diagram is often drawn as follows:

It certainly looks a lot simpler. Why do you suppose I opted for the more complicated version?

Proposition 3.3.3. *If two objects in some concrete category are free on the same set, then they are isomorphic.*

Definition 3.3.4. Let A be a subset of a nonempty set S and $\mathbb{F}$ be a field. Define $\chi_A\colon S \to \mathbb{F}$, the CHARACTERISTIC FUNCTION of A, by

$$\chi_A(x) = \begin{cases} 1, & \text{if } x \in A \\ 0, & \text{otherwise} \end{cases}$$

Example 3.3.5. If S is an arbitrary nonempty set and $\mathbb{F}$ is a field, then there exists a vector space V over $\mathbb{F}$ that is free on S. This vector space is unique (up to isomorphism).

Hint for proof. Given the set S let V be the set of all $\mathbb{F}$-valued functions on S that have finite support. Define addition and scalar multiplication pointwise. The map $\iota\colon s \mapsto \chi_{\{s\}}$ of each element $s \in S$ to the characteristic function of $\{s\}$ is the desired injection. To verify that V is free over S it must be shown that for every vector space W and every function $S \xrightarrow{\;f\;} |W|$ there exists a unique linear map $V \xrightarrow{\;\tilde{f}\;} W$ which makes the following diagram commute.

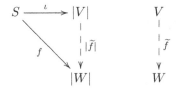

Proposition 3.3.6. *Every vector space is free.*

Hint for proof. Of course, part of the problem is to specify a set S on which the given vector space is free.

Exercise 3.3.7. Let $S = \{a, *, \#\}$. Then an expression such as

$$3a - \tfrac{1}{2} * + \sqrt{2}\,\#$$

is said to be a *formal linear combination* of elements of S. Make sense of such expressions.

3.4. Products and Coproducts

In this section we define products and coproducts of vector spaces in terms of universal mapping properties.

Definition 3.4.1. Let A_1 and A_2 be objects in a category $\mathbf{C}$. We say that a triple (P, π_1, π_2), where P is an object and $\pi_k\colon P \to A_k$ ($k = 1, 2$) are morphisms, is a PRODUCT of A_1 and A_2 if for every object B in $\mathbf{C}$ and every pair of morphisms $f_k\colon B \to A_k$ ($k = 1, 2$) there exists a unique map

$g\colon B \to P$ such that $f_k = \pi_k \circ g$ for $k = 1,\ 2$.

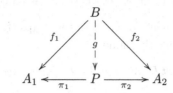

A triple (P, j_1, j_2), where P is an object and $j_k\colon A_k \to P$, $(k = 1,\ 2)$ are morphisms, is a COPRODUCT of A_1 and A_2 if for every object B in $\mathbf{C}$ and every pair of morphisms $F_k\colon A_k \to B$ $(k = 1,\ 2)$ there exists a unique map $G\colon P \to B$ such that $F_k = G \circ j_k$ for $k = 1,\ 2$.

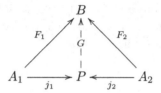

Proposition 3.4.2. *In an arbitrary category products and coproducts (if they exist) are essentially unique.*

"Essentially unique" means unique up to isomorphism. Thus in the preceding proposition the claim is that if (P, π_1, π_2) and (Q, ρ_1, ρ_2) are both products of two given objects, then $P \cong Q$.

Definition 3.4.3. Let V and W be vector spaces over the same field $\mathbb{F}$. To make the Cartesian product $V \times W$ into a vector space we define addition by

$$(v, w) + (v', w') = (v + v', w + w')$$

(where $v,\ v' \in V$ and $w,\ w' \in W$), and we define scalar multiplication by

$$\alpha(v, w) = (\alpha v, \alpha w)$$

(where $\alpha \in \mathbb{F}$, $v \in V$, and $w \in W$). The resulting vector space we call the (EXTERNAL) DIRECT SUM of V and W. It is conventional to use the same notation $V \oplus W$ for external direct sums that we use for internal direct sums.

Example 3.4.4. The external direct sum of two vector spaces (as defined in 3.4.3) is a vector space.

Example 3.4.5. In the category of vector spaces and linear maps the external direct sum is not only a product but also a coproduct.

Example 3.4.6. In the category of sets and maps (functions) the product and the coproduct are *not* the same.

Proposition 3.4.7. *Let U, V, and W be vector spaces. If $U \cong W$, then $U \oplus V \cong W \oplus V$.*

Example 3.4.8. The converse of the preceding proposition is not true.

Definition 3.4.9. Let V_0, V_1, V_2, ... be vector spaces (over the same field). Then their (EXTERNAL) DIRECT SUM, which is denoted by $\bigoplus_{k=0}^{\infty} V_k$, is defined to be the set of all functions $v \colon \mathbb{Z}^+ \to \bigcup_{k=0}^{\infty} V_k$ with finite support such that $v(k) = v_k \in V_k$ for each $k \in \mathbb{Z}^+$. The usual pointwise addition and scalar multiplication make this set into a vector space.

3.5. Quotients

Definition 3.5.1. Let A be an object in a concrete category $\mathbf{C}$. A surjective morphism $A \xrightarrow{\pi} B$ in $\mathbf{C}$ is a QUOTIENT MAP for A if a function $g \colon B \to C$ (in **SET**) is a morphism (in **C**) whenever $g \circ \pi$ is a morphism. An object B in $\mathbf{C}$ is a QUOTIENT OBJECT for A if it is the range of some quotient map for A.

Proposition 3.5.2. *In the category of vector spaces and linear maps every surjective linear map is a quotient map.*

The next item shows how a particular quotient object can be generated by "factoring out a subspace".

Definition 3.5.3. Let M be a subspace of a vector space V. Define an equivalence relation $\sim$ on V by

$$x \sim y \quad \text{if and only if} \quad y - x \in M.$$

For each $x \in V$ let $[x]$ be the equivalence class containing x. Let V/M be the set of all equivalence classes of elements of V. For $[x]$ and $[y]$ in V/M define

$$[x] + [y] := [x + y]$$

and for $\alpha \in \mathbb{R}$ and $[x] \in V/M$ define

$$\alpha[x] := [\alpha x].$$

Under these operations V/M becomes a vector space. It is the QUOTIENT SPACE of V by M. The notation V/M is usually read "V mod M". The linear map

$$\pi: V \to V/M: x \mapsto [x]$$

is called the QUOTIENT MAP.

Exercise 3.5.4. Verify the assertions made in Definition 3.5.3. In particular, show that $\sim$ is an equivalence relation, that addition and scalar multiplication of the set of equivalence classes is well defined, that under these operations V/M is a vector space, and that the quotient map is linear.

The following result is called the *fundamental quotient theorem* or the first isomorphism theorem for vector spaces.

Theorem 3.5.5. *Let V and W be vector spaces and $M \preceq V$. If $T \in \mathcal{L}(V, W)$ and $\ker T \supseteq M$, then there exists a unique $\widetilde{T} \in \mathcal{L}(V/M, W)$ that makes the following diagram commute.*

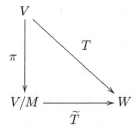

Furthermore, $\widetilde{T}$ is injective if and only if $\ker T = M$; and $\widetilde{T}$ is surjective if and only if T is.

Corollary 3.5.6. *If $T: V \to W$ is a linear map between vector spaces, then $\operatorname{ran} T \cong V/\ker T$.*

For obvious reasons the next result is usually called the *rank-plus-nullity theorem.* (It is also sometimes listed as part of the *fundamental theorem of linear algebra.*)

Proposition 3.5.7. *Let $T: V \to W$ be a linear map between vector spaces. If V is finite dimensional, then*

$$\operatorname{rank} T + \operatorname{nullity} T = \dim V.$$

Hint for proof. Let M be any subspace complementary to $\ker T$ in V. Show that the restriction of T to M establishes an isomorphism from M to $\operatorname{ran} T$.

Corollary 3.5.8. *If M is a subspace of a finite dimensional vector space V, then* $\dim V/M = \dim V - \dim M$.

3.6. Exact Sequences

Definition 3.6.1. A sequence of vector spaces and linear maps

$$\cdots \longrightarrow V_{n-1} \xrightarrow{\ j_n\ } V_n \xrightarrow{\ j_{n+1}\ } V_{n+1} \longrightarrow \cdots$$

is said to be EXACT AT V_n if $\operatorname{ran} j_n = \ker j_{n+1}$. A sequence is EXACT if it is exact at each of its constituent vector spaces. A sequence of vector spaces and linear maps of the form

$$0 \longrightarrow U \xrightarrow{\ j\ } V \xrightarrow{\ k\ } W \longrightarrow 0$$

is a SHORT EXACT SEQUENCE. (Here $\mathbf{0}$ denotes the trivial 0-dimensional vector space, and the unlabeled arrows are the obvious linear maps.)

Proposition 3.6.2. *The sequence*

$$0 \longrightarrow U \xrightarrow{\ j\ } V \xrightarrow{\ k\ } W \longrightarrow 0$$

of vector spaces is exact at U if and only if j is injective. It is exact at W if and only if k is surjective.

Exercise 3.6.3. Suppose $a < b$. Let $\mathcal{K}$ be the family of constant functions on the interval $[a, b]$, $\mathcal{C}^1$ be the family of all continuously differentiable functions on $[a, b]$, and $\mathcal{C}$ be the family of all continuous functions on $[a, b]$. (A function f is said to be CONTINUOUSLY DIFFERENTIABLE if its derivative f' exists and is continuous.)

Specify linear maps j and k so that the following sequence is short exact:

$$0 \longrightarrow \mathcal{K} \xrightarrow{\ j\ } \mathcal{C}^1 \xrightarrow{\ k\ } \mathcal{C} \longrightarrow 0.$$

Exercise 3.6.4. Let $\mathcal{C}$ be the family of all continuous functions on the interval $[0, 2]$. Let E_1 be the mapping from $\mathcal{C}$ into $\mathbb{R}$ defined by $E_1(f) = f(1)$. (The functional E_1 is called *evaluation at 1*.)

Find a subspace $\mathcal{F}$ of C such that the following sequence is short exact.

$$0 \longrightarrow \mathcal{F} \overset{\iota}{\longrightarrow} C \overset{E_1}{\longrightarrow} \mathbb{R} \longrightarrow 0.$$

Exercise 3.6.5. If $j \colon U \to V$ is an injective linear map between vector spaces, then the sequence

$$0 \longrightarrow U \overset{j}{\longrightarrow} V \overset{\pi}{\longrightarrow} V/\operatorname{ran} j \longrightarrow 0$$

is exact.

Example 3.6.6. Let U and V be vector spaces. Then the following sequence is short exact:

$$0 \longrightarrow U \overset{\iota_1}{\longrightarrow} U \oplus V \overset{\pi_2}{\longrightarrow} V \longrightarrow 0.$$

The indicated linear maps are the obvious ones:

$$\iota_1 \colon U \to U \oplus V \colon u \mapsto (u, 0)$$

and

$$\pi_2 \colon U \oplus V \to V \colon (u, v) \mapsto v.$$

Proposition 3.6.7. *Consider the following diagram in the category of vector spaces and linear maps.*

$$
\begin{array}{ccccccccc}
0 & \longrightarrow & U & \overset{j}{\longrightarrow} & V & \overset{k}{\longrightarrow} & W & \longrightarrow & 0 \\
& & \downarrow{\scriptstyle f} & & \downarrow{\scriptstyle g} & & \vdots{\scriptstyle h} & & \\
0 & \longrightarrow & U' & \underset{j'}{\longrightarrow} & V' & \underset{k'}{\longrightarrow} & W' & \longrightarrow & 0
\end{array}
$$

If the rows are exact and the left square commutes, then there exists a unique linear map $h \colon W \to W'$ that makes the right square commute.

Proposition 3.6.8 (The Short Five Lemma). *Consider the following diagram of vector spaces and linear maps*

$$
\begin{array}{ccccccccc}
0 & \longrightarrow & U & \overset{j}{\longrightarrow} & V & \overset{k}{\longrightarrow} & W & \longrightarrow & 0 \\
& & \downarrow{\scriptstyle f} & & \downarrow{\scriptstyle g} & & \downarrow{\scriptstyle h} & & \\
0 & \longrightarrow & U' & \underset{j'}{\longrightarrow} & V' & \underset{k'}{\longrightarrow} & W' & \longrightarrow & 0
\end{array}
$$

where the rows are exact and the squares commute. Then the following hold.

(a) If g is surjective, so is h.
(b) If f is surjective and g is injective, then h is injective.
(c) If f and h are surjective, so is g.
(d) If f and h are injective, so is g.

Proposition 3.6.9. *Show that if* $0 \longrightarrow U \xrightarrow{\ j\ } V \xrightarrow{\ k\ } W \longrightarrow 0$ *is an exact sequence of vector spaces and linear maps, then* $V \cong U \oplus W$.

Hint for proof. Consider the following diagram and use Proposition 3.6.8.

$$
\begin{array}{ccccccccc}
0 & \longrightarrow & U & \xrightarrow{\ j\ } & V & \xrightarrow{\ k\ } & W & \longrightarrow & 0 \\
  &   & \| & & \downarrow{\scriptstyle g} & & \| & & \\
0 & \longrightarrow & U & \xrightarrow[i_1]{} & U \oplus W & \underset{i_2}{\overset{\pi_2}{\rightleftarrows}} & W & \longrightarrow & 0
\end{array}
$$

The trick is to find the right map g.

Exercise 3.6.10. Prove the converse of the preceding exercise. That is, suppose that U, V, and W are vector spaces and that $V \cong U \oplus W$; prove that there exist linear maps j and k such that the sequence $0 \longrightarrow U \xrightarrow{\ j\ } V \xrightarrow{\ k\ } W \longrightarrow 0$ is exact. *Hint.* Suppose $g \colon U \oplus W \to V$ is an isomorphism. Define j and k in terms of g.

Proposition 3.6.11. *Let M and M' be subspaces of vector spaces V and V', respectively. Prove the following.*

(a) *Every linear map $T \colon V \to V'$ that satisfies $T(M) \subseteq M'$ induces a linear map $\widehat{T}$ from V/M into V'/M' that satisfies $\widehat{T}(v + M) = (Tv) + M'$ for all $v \in V$.*
(b) *If T is an isomorphism and $T(M) = M'$, then $V/M \cong V'/M'$.*

Proposition 3.6.12. *If* $0 \longrightarrow U \xrightarrow{\ j\ } V \xrightarrow{\ k\ } W \longrightarrow 0$ *is an exact sequence of vector spaces and linear maps, then* $W \cong V/\operatorname{ran} j$. *Thus, if $U \preceq V$ and j is the inclusion map, then* $W \cong V/U$.

Give two different proofs of the preceding result: one using Theorem 3.5.5 and the other using Proposition 3.6.13.

Proposition 3.6.13. *The converse of Proposition 3.6.12 is also true. That is, if $j \colon U \to V$ is an injective linear map between vector spaces and*

$W \cong V/\operatorname{ran} j$, *then there exists a linear map k that makes the sequence*

$$0 \longrightarrow U \xrightarrow{\ j\ } V \xrightarrow{\ k\ } W \longrightarrow 0 \ exact.$$

Proposition 3.6.14. *If $V_0, V_1, \ldots, V_n$ are finite dimensional vector spaces and the sequence*

$$0 \longrightarrow V_n \xrightarrow{\ d_n\ } V_{n-1} \longrightarrow \cdots \longrightarrow V_1 \xrightarrow{\ d_1\ } V_0 \longrightarrow 0$$

is exact, then $\sum_{k=0}^{n} (-1)^k \dim V_k = 0$.

3.7. Some Miscellaneous Results

Definition 3.7.1. Let $T \in \mathcal{L}(V, W)$, where V and W are vector spaces. Define $\operatorname{coker} T$, the COKERNEL of T, to be $W/\operatorname{ran} T$.

Proposition 3.7.2. *Let U, V, and W be vector spaces. If $S \in \mathcal{L}(U, V)$, $T \in \mathcal{L}(V, W)$, then the sequence*

$$0 \longrightarrow \ker S \longrightarrow \ker TS \longrightarrow \ker T \longrightarrow \operatorname{coker} S \longrightarrow \operatorname{coker} TS \longrightarrow \operatorname{coker} T \longrightarrow 0$$

is exact.

Proposition 3.7.3. *Let W be a vector space and $M \preceq V \preceq W$. Then*

$$(W/M)/(V/M) \cong W/V.$$

Hint for proof. Proposition 3.6.12.

Proposition 3.7.4. *Let V be a vector space and M, $M' \preceq V$ Then*

$$(M + M')/M \cong M'/(M \cap M').$$

Proposition 3.7.5. *Let M be a subspace of a vector space V. Then the following are equivalent:*

(a) $\dim V/M < \infty$;
(b) *there exists a finite dimensional subspace F of V such that $V = M \oplus F$; and*
(c) *there exists a finite dimensional subspace F of V such that $V = M + F$.*

Exercise 3.7.6. Suppose that a vector space V is the direct sum of subspaces U and W. Some authors define the CODIMENSION of U to be $\dim W$. Others define it to be $\dim V/U$. Show that these are equivalent.

Chapter 4

THE SPECTRAL THEOREM FOR VECTOR SPACES

4.1. Projections

Much of mathematical research consists analyzing complex objects by writing them as a combination of simpler objects. In the case of vector space operators the simpler objects, the fundamental building blocks, are *projection operators*.

Definition 4.1.1. Let V be a vector space. An operator $E \in \mathfrak{L}(V)$ is a PROJECTION OPERATOR if it is IDEMPOTENT; that is, if $E^2 = E$.

Proposition 4.1.2. *If E is a projection operator on a vector space V, then*

$$V = \operatorname{ran} E \oplus \ker E.$$

Proposition 4.1.3. *Let V be a vector space and E, $F \in \mathfrak{L}(V)$. If $E + F = I_V$ and $EF = 0$, then E and F are projection operators and $V = \operatorname{ran} E \oplus \operatorname{ran} F$.*

Proposition 4.1.4. *Let V be a vector space and $E_1, \ldots, E_n \in \mathfrak{L}(V)$. If $\sum_{k=1}^{n} E_k = I_V$ and $E_i E_j = 0$ whenever $i \neq j$, then each E_k is a projection operator and $V = \bigoplus_{k=1}^{n} \operatorname{ran} E_k$.*

Proposition 4.1.5. *If E is a projection operator on a vector space V, then $\operatorname{ran} E = \{x \in V \colon Ex = x\}$.*

Proposition 4.1.6. *Let E and F be projection operators on a vector space V. Then $E + F = I_V$ if and only if $EF = FE = 0$ and $\ker E = \operatorname{ran} F$.*

Definition 4.1.7. Let V be a vector space and suppose that $V = M \oplus N$. We know from an earlier Theorem 1.5.20 that for each $v \in V$ there exist unique vectors $m \in M$ and $n \in N$ such that $v = m + n$. Define a function $E_{NM} : V \to V$ by $E_{NM} v = m$. The function E_{NM} is called the PROJECTION OF V ALONG N ONTO M. (This terminology is, of course, optimistic. We must *prove* that E_{NM} is in fact a projection operator.)

Proposition 4.1.8. *If $M \oplus N$ is a direct sum decomposition of a vector space V, then the function E_{NM} defined in 4.1.7 is a projection operator whose range is M and whose kernel is N.*

Proposition 4.1.9. *If $M \oplus N$ is a direct sum decomposition of a vector space V, then $E_{NM} + E_{MN} = I_V$ and $E_{NM} E_{MN} = \mathbf{0}$.*

Proposition 4.1.10. *If E is a projection operator on a vector space V, then there exist M, $N \preccurlyeq V$ such that $E = E_{NM}$.*

Exercise 4.1.11. Let M be the line $y = 2x$ and N be the y-axis in $\mathbb{R}^2$. Find $[E_{MN}]$ and $[E_{NM}]$.

Exercise 4.1.12. Let E be the projection of $\mathbb{R}^3$ onto the plane $3x - y + 2z = 0$ along the z-axis. Find the matrix representation $[E]$ (of E with respect to the standard basis of $\mathbb{R}^3$).

Exercise 4.1.13. Let F be the projection of $\mathbb{R}^3$ onto the z-axis along the plane $3x - y + 2z = 0$. Where does F take the point $(4, 5, 1)$?

Exercise 4.1.14. Let P be the plane in $\mathbb{R}^3$ whose equation is $x + 2y - z = 0$ and L be the line whose equations are $\dfrac{x}{3} = y = \dfrac{z}{2}$. Let E be the projection of $\mathbb{R}^3$ along L onto P and F be the projection of $\mathbb{R}^3$ along P onto L. Then

$$[E] = \frac{1}{3} \begin{bmatrix} a & -b & c \\ -d & d & d \\ a - 2d & -b + 2d & c + 2d \end{bmatrix} \quad \text{and} \quad [F] = \frac{1}{3} \begin{bmatrix} 3d & 3e & -3d \\ d & e & -d \\ 2d & 2e & -2d \end{bmatrix}$$

where $a = \underline{\quad}$, $b = \underline{\quad}$, $c = \underline{\quad}$, $d = \underline{\quad}$, and $e = \underline{\quad}$.

Exercise 4.1.15. Let $T \colon V \to W$ be linear and $S \colon W \to V$ a left inverse for T. Then

(a) $W = \operatorname{ran} T \oplus \ker S$, and

(b) TS is the projection along $\ker S$ onto $\operatorname{ran} T$.

4.2. Algebras

Definition 4.2.1. Let $(A, +, M)$ be a vector space over a field $\mathbb{F}$ which is equipped with another binary operation $\cdot\colon A \times A \to A\colon (a, b) \mapsto ab$ in such a way that $(A, +, \cdot)$ is a ring. If additionally the equations

$$\alpha(ab) = (\alpha a)b = a(\alpha b) \qquad (4.1)$$

hold for all a, $b \in A$ and $\alpha \in \mathbb{F}$, then $(A, +, M, \cdot)$ is an ALGEBRA over the field $\mathbb{F}$ (sometimes referred to as a LINEAR ASSOCIATIVE ALGEBRA). We abuse notation in the usual way by writing such things as, "Let A be an algebra". We say that an algebra A is UNITAL if its underlying ring $(A, +, \cdot)$ is. And it is COMMUTATIVE if its ring is.

Example 4.2.2. A field may be regarded as an algebra over itself.

Example 4.2.3. If S is a nonempty set, then the vector space $\mathcal{F}(S, \mathbb{F})$ (see Example 1.4.5) is a commutative unital algebra under POINTWISE MULTI-PLICATION, which is defined for all f, $g \in \mathcal{F}(S, \mathbb{F})$ by

$$(f \cdot g)(s) = f(s) \cdot g(s)$$

for all $s \in S$. The constant function $\mathbf{1}$ (that is, the function whose value at each $s \in S$ is 1) is the multiplicative identity.

Example 4.2.4. If V is a vector space, then the set $\mathfrak{L}(V)$ of linear operators on V is a unital algebra under pointwise addition, pointwise scalar multiplication, and composition.

Notation 4.2.5. In the following material we make the notational convention that if B and C are subsets of (a ring or) an algebra A, then BC denotes the set of all sums of products of elements in B and C. That is

$$BC := \{b_1c_1 + \cdots + b_nc_n \colon n \in \mathbb{N};\ b_1, \ldots, b_n \in B;\ \text{and}\ c_1, \ldots, c_n \in C\}.$$

And, of course, if $b \in A$, then $bC = \{b\}C$.

Definition 4.2.6. A map $f\colon A \to B$ between algebras is an (ALGE-BRA) HOMOMORPHISM if it is a linear map between A and B as vector spaces that preserves multiplication (in the sense of Equation (1.2) in Section 1.3). In other words, an algebra homomorphism is a linear ring homomorphism. It is a UNITAL (ALGEBRA) HOMOMORPHISM if it preserves identities (as in Equation (1.3) in Section 1.3). The KERNEL of an algebra homomorphism $f\colon A \to B$ is, of course, $\{a \in A \colon f(a) = \mathbf{0}\}$.

If f^{-1} exists and is also an algebra homomorphism, then f is an ISO-MORPHISM from A to B. If an isomorphism from A to B exists, then A and B are ISOMORPHIC.

Here are three essentially obvious facts about algebra homomorphisms.

Proposition 4.2.7. *Every bijective algebra (or ring) homomorphism is an isomorphism.*

Proposition 4.2.8. *If $f\colon A \to B$ is an isomorphism between algebras (or rings) and A is unital, then so is B and f is a unital homomorphism.*

Proposition 4.2.9. *Let A, B, and C be algebras (or rings). If $f\colon A \to B$ and $g\colon B \to C$ are homomorphisms, so is $gf\colon A \to C$. (As is the case with group homomorphism and linear maps, gf denotes the composite function $g \circ f$.) If f and g are unital, so is gf.*

Here is an example of another important algebra.

Example 4.2.10. We have seen in Example 1.4.8 that the set $\mathbf{M}_n$ of $n \times n$ matrices of real numbers is a vector space. If $a = \begin{bmatrix} a_{ij} \end{bmatrix}$ and $b = \begin{bmatrix} b_{kl} \end{bmatrix}$ are $n \times n$ matrices of real numbers, then the PRODUCT of a and b is the $n \times n$ matrix $c = ab$ whose entry in the ith row and k^{th} column is $c_{ik} = \sum_{j=1}^{n} a_{ij} b_{jk}$. This definition makes $\mathbf{M}_n$ into a unital algebra.

Hint for proof. Proving associativity of matrix multiplication can be something of a nuisance if one charges ahead without thinking. As an alternative to brute calculation look at Exercise 2.2.16.

Definition 4.2.11. A subset of an algebra A that is closed under the operations of addition, multiplication, and scalar multiplication is a SUB-ALGEBRA of A. If A is a unital algebra and B is a subalgebra of A that contains the multiplicative identity of A, then B is a UNITAL SUBALGEBRA of A.

Caution 4.2.12. Be very careful with the preceding definition. It is possible for B to be a subalgebra of an algebra A and to be a unital algebra but still not be a unital subalgebra of A! The definition requires that for B to be a unital subalgebra of A the identity of B must be the same as the identity of A. *Example:* Under pointwise operations $A = \mathbb{R}^2$ is a unital algebra. The set $B = \{(x, 0) \colon x \in \mathbb{R}\}$ is a subalgebra of A. And certainly

B is unital (the element $(1,0)$ is the multiplicative identity of B). But B is *not* a unital subalgebra of A because it does not contain the multiplicative identity $(1,1)$ of A.

Example 4.2.13. Let S be a nonempty set. The family $\mathcal{B}(S)$ of all bounded real valued functions on S is a unital subalgebra of the algebra $\mathcal{F}(S)$ of all real valued functions on S.

Definition 4.2.14. A LEFT IDEAL in an algebra A is a vector subspace J of A such that $AJ \subseteq J$. (For RIGHT IDEALS, of course, we require $JA \subseteq J$.) We say that J is an IDEAL if it is a two-sided ideal; that is, both a left and a right ideal. A PROPER ideal is an ideal that is a proper subset of A.

The ideals $\{0\}$ and A are often referred to as the TRIVIAL IDEALS of A. The algebra A is SIMPLE if it has no nontrivial ideals.

Example 4.2.15. If $\phi \colon A \to B$ is an algebra homomorphism, then the kernel of ϕ is an ideal in A and the range of ϕ is a subalgebra of B.

Definition 4.2.16. An element a of a unital algebra A is INVERTIBLE if there exists an element $a^{-1} \in A$ such that $aa^{-1} = a^{-1}a = 1_A$.

Proposition 4.2.17. *If a and b are invertible elements in a unital algebra, then ab is also invertible and $(ab)^{-1} = b^{-1}a^{-1}$.*

Proposition 4.2.18. *No invertible element in a unital algebra can belong to a proper ideal.*

Proposition 4.2.19. *Let a be an element of a commutative algebra A. Then aA is an ideal in A. If A is unital and a is not invertible, then aA is a proper ideal in A.*

Example 4.2.20. Let $\mathfrak{J}$ be a family of ideals in an algebra A. Then the $\bigcap \mathfrak{J}$ is an ideal in A.

Definition 4.2.21. Let a be an element of an algebra A. Then the intersection of all the ideals of A that contain a is the PRINCIPAL IDEAL generated by a.

Proposition 4.2.22. *Let a be an element of a commutative algebra A. The ideal aA in Proposition 4.2.19 is the principal ideal generated by a.*

4.3. Quotients and Unitizations

Definition 4.3.1. Let J be a proper ideal in an algebra A. Define an equivalence relation $\sim$ on A by

$$a \sim b \quad \text{if and only if} \quad b - a \in J.$$

For each $a \in A$ let $[a]$ be the equivalence class containing a. Let A/J be the set of all equivalence classes of elements of A. For $[a]$ and $[b]$ in A/J define

$$[a] + [b] := [a + b] \quad \text{and} \quad [a][b] := [ab]$$

and for $\alpha \in \mathbb{C}$ and $[a] \in A/J$ define

$$\alpha[a] := [\alpha a].$$

Under these operations A/J becomes an algebra. It is the QUOTIENT ALGEBRA of A by J. The notation A/J is usually read "A mod J". The surjective algebra homomorphism

$$\pi \colon A \to A/J \colon a \mapsto [a]$$

is called the QUOTIENT MAP.

Exercise 4.3.2. Verify the assertions made in the preceding definition.

Definition 4.3.3. Let A be an algebra over a field $\mathbb{F}$. The UNITIZATION of A is the unital algebra $\tilde{A} = A \times \mathbb{F}$ in which addition and scalar multiplication are defined pointwise and multiplication is defined by

$$(a, \lambda) \cdot (b, \mu) = (ab + \mu a + \lambda b, \lambda\mu).$$

Exercise 4.3.4. Prove that the unitization $\tilde{A}$ of an algebra A is in fact a unital algebra with $(0, 1)$ as its identity. Prove also that A is (isomorphic to) a subalgebra of $\tilde{A}$ with codimension 1.

4.4. The Spectrum

Definition 4.4.1. Let a be an element of a unital algebra A over a field $\mathbb{F}$. The SPECTRUM of a, denoted by $\sigma_A(a)$ or just $\sigma(a)$, is the set of all $\lambda \in \mathbb{F}$ such that $a - \lambda\mathbf{1}$ is not invertible.

If the algebra A is not unital we will still speak of *the spectrum* of the element a with the understanding that we are speaking of the spectrum of a in the unitization of A.

Example 4.4.2. If z is an element of the algebra $\mathbb{C}$ of complex numbers, then $\sigma(z) = \{z\}$.

Example 4.4.3. Let f be an element of the algebra $\mathcal{C}([a, b])$ of continuous complex valued functions on the interval $[a, b]$. Then the spectrum of f is its range.

Example 4.4.4. The operator that rotates (the real vector space) $\mathbb{R}^2$ by $\frac{\pi}{2}$ radians has empty spectrum.

For the next example you may assume that a square matrix of real or complex numbers is invertible if and only if its determinant is nonzero.

Example 4.4.5. The family $\mathbf{M}_3(\mathbb{C})$ of 3×3 matrices of complex numbers is a unital algebra under the usual matrix operations. The spectrum of the

matrix $\begin{bmatrix} 5 & -6 & -6 \\ -1 & 4 & 2 \\ 3 & -6 & -4 \end{bmatrix}$ is $\{1, 2\}$.

Example 4.4.6. Let a be an element of a unital complex algebra such that $a^2 = \mathbf{1}$. Then either

(i) $a = \mathbf{1}$, in which case $\sigma(a) = \{1\}$, or
(ii) $a = -\mathbf{1}$, in which case $\sigma(a) = \{-1\}$, or
(iii) $\sigma(a) = \{-1, 1\}$.

Hint for proof. In (iii) to prove $\sigma(a) \subseteq \{-1, 1\}$, consider $\dfrac{1}{1 - \lambda^2}(a + \lambda \mathbf{1})$.

Definition 4.4.7. An element a of an algebra is IDEMPOTENT if $a^2 = a$.

Example 4.4.8. Let a be an idempotent element of a unital complex algebra. Then either

(i) $a = \mathbf{1}$, in which case $\sigma(a) = \{1\}$, or
(ii) $a = \mathbf{0}$, in which case $\sigma(a) = \{0\}$, or
(iii) $\sigma(a) = \{0, 1\}$.

Hint for proof. In (iii) to prove $\sigma(a) \subseteq \{0, 1\}$, consider $\dfrac{1}{\lambda - \lambda^2}(a + (\lambda - 1)\mathbf{1})$.

4.5. Polynomials

Notation 4.5.1. If S is a set and A is an algebra, $l(S, A)$ denotes the vector space of all functions from S into A with pointwise operations of addition and scalar multiplication, and $l_c(S, A)$ denotes the subspace of functions with finite support.

Definition 4.5.2. Let A be a unital commutative algebra. On the vector space $l(\mathbb{Z}^+, A)$ define a binary operation $*$ (often called CONVOLUTION) by $(f * g)_n = \sum_{j+k=n} f_j\, g_k = \sum_{j=0}^{n} f_j\, g_{n-j}$ (where f, $g \in l(\mathbb{Z}^+, A)$ and $n \in \mathbb{Z}^+$). An element of $l(\mathbb{Z}^+, A)$ is a FORMAL POWER SERIES (with coefficients in A) and an element of $l_c(\mathbb{Z}^+, A)$ is a POLYNOMIAL (with coefficients in A). The reason for using the expression "with *coefficients* in A" may seem at first a bit mysterious. For an explanation look at Example 4.5.10.

Proposition 4.5.3. *If A is a unital commutative algebra, then, under the operations defined in 4.5.2, $l(\mathbb{Z}^+, A)$ is a unital commutative algebra (whose multiplicative identity is the sequence $(1_A, 0, 0, 0, \ldots)$) and $l_c(\mathbb{Z}^+, A)$ is a unital subalgebra of $l(\mathbb{Z}^+, A)$.*

Proposition 4.5.4. *If $\phi\colon A \to B$ is a unital algebra homomorphism between unital commutative algebras, then the map*

$$l(\mathbb{Z}^+, \phi)\colon l(\mathbb{Z}^+, A) \to l(\mathbb{Z}^+, B)\colon f \mapsto \big(\phi(f_n)\big)_{n=0}^{\infty}$$

is also a unital homomorphism of unital commutative algebras. The pair of maps $A \mapsto l(\mathbb{Z}^+, A)$ and $\phi \mapsto l(\mathbb{Z}^+, \phi)$ is a covariant functor from the category of unital commutative algebras and unital algebra homomorphisms to itself.

Remark 4.5.5. We regard the algebra A as a subset of $l(\mathbb{Z}^+, A)$ by identifying the element $a \in A$ with the element $(a, 0, 0, 0, \ldots) \in l(\mathbb{Z}^+, A)$. Thus the map $a \mapsto (a, 0, 0, 0, \ldots)$ becomes an inclusion map. (Technically speaking, of course, the map $\psi\colon a \mapsto (a, 0, 0, 0, \ldots)$ is an injective unital homomorphism and $A \cong \operatorname{ran} \psi$.)

Convention 4.5.6. In the algebra $l(\mathbb{Z}^+, A)$ we will normally write $a \cdot b$ or ab for $a * b$.

Definition 4.5.7. Let A be a unital commutative algebra. In the algebra $l(\mathbb{Z}^+, A)$ of formal power series the special sequence $x = (0, 1_A, 0, 0, 0, \ldots)$ is called the INDETERMINANT of $l(\mathbb{Z}^+, A)$. Notice that the sequence $x^2 = x \cdot x = (0, 0, 1, 0, 0, 0, \ldots)$, the sequence $x^3 = x \cdot x \cdot x = (0, 0, 0, 1, 0, 0, 0, \ldots)$, and so on. It is conventional to take x^0 to be the multiplicative identity $(1_A, 0, 0, 0, \ldots)$ in $l(\mathbb{Z}^+, A)$.

Remark 4.5.8. The algebra $l(\mathbb{Z}^+, A)$ of formal power series with coefficients in a unital commutative algebra A is frequently denoted by $A[[x]]$ and the subalgebra $l_c(\mathbb{Z}^+, A)$ of polynomials is denoted by $A[x]$.

For many algebraists scalar multiplication is of little interest so A is taken to be a unital commutative ring, so that $A[[x]]$ is a *ring* of formal power series (with coefficients in A) and $A[x]$ is the *polynomial ring* (with coefficients in A). We will be primarily interested in the case where A is a field $\mathbb{F}$. Since a field can be regarded as a one-dimensional vector space over itself, it is also an algebra. Thus we will take $\mathbb{F}[x]$ to be the *polynomial algebra* with coefficients in $\mathbb{F}$; it has as its basis $\{x^n \colon n = 0, 1, 2, \ldots\}$.

Definition 4.5.9. A nonzero polynomial p, being an element of $l_c(\mathbb{Z}^+, A)$, has finite support. So there exists $n_0 \in \mathbb{Z}^+$ such that $p_n = 0$ whenever $n > n_0$. The smallest such n_0 is the DEGREE of the polynomial. We denote it by $\deg p$. A polynomial of degree 0 is a CONSTANT POLYNOMIAL. The zero polynomial (the additive identity of $l(\mathbb{Z}^+, A)$) is a special case; while it is also a constant polynomial some authors assign it no degree whatever, while others let its degree be $-\infty$.

If p is a polynomial of degree n, then p_n is the LEADING COEFFICIENT of p. A polynomial is MONIC if its leading coefficient is 1.

Example 4.5.10. Let A be a unital commutative algebra. If p is a nonzero polynomial in $l_c(\mathbb{Z}^+, A)$, then

$$p = \sum_{k=0}^{n} p_k x^k \quad \text{where } n = \deg p.$$

This is the STANDARD FORM of the polynomial p. Keep in mind that each coefficient p_k belongs to the algebra A. Also notice that it does not really matter whether we write p as $\sum_{k=0}^{n} p_k x^k$ or as $\sum_{k=0}^{\infty} p_k x^k$; so frequently we write just $\sum p_k x^k$.

Remark 4.5.11. Recall that there is occasionally a slight ambiguity in notation for sets. For example, if we consider the (complex) solutions to an algebraic equation E of degree n, we know that, *counting multiplicities*, there are n solutions to the equation. So it is common practice to write, "Let $\{x_1, x_2, \ldots, x_n\}$ be the set of solutions to E". Notice that in this context there may be repeated elements of the set. The cardinality of the set may be strictly less than n. However, when we encounter the expression, "Let $\{x_1, x_2, \ldots, x_n\}$ be a set of $\ldots$," it is usually the intention of the author that the elements are distinct, that the cardinality of the set is n.

A similar ambiguity arises in polynomial notation. If, for example, $p = \sum_{k=0}^{n} p_k x^k$ and $q = \sum_{k=0}^{n} q_k x^k$ are both polynomials of degree n, we ordinarily write their sum as $p + q = \sum_{k=0}^{n} (p_k + q_k) x^k$ even though the resulting sum may very well have degree strictly less than n. On the other hand when one sees, "Consider a polynomial $p = \sum_{k=0}^{n} p_k x^k$ such that . . . ," it is usually intended that p have degree n; that is, that p is written in standard form.

Proposition 4.5.12. *If p and q are polynomials with coefficients in a unital commutative algebra A, then*

(i) $\deg(p + q) \leq \max\{\deg p, \deg q\}$, *and*
(ii) $\deg(pq) \leq \deg p + \deg q$.

If A is a field, then equality holds in (ii).

Example 4.5.13. If A is a unital commutative algebra, then so is $l(A, A)$ under pointwise operations of addition, multiplication, and scalar multiplication.

Definition 4.5.14. Let A be a unital commutative algebra over a field $\mathbb{F}$. For each polynomial $p = \sum_{k=0}^{n} p_k x^k$ with coefficients in $\mathbb{F}$ define

$$\widetilde{p} \colon A \to A \colon a \mapsto \sum_{k=0}^{n} p_k a^k.$$

Then $\widetilde{p}$ is the POLYNOMIAL FUNCTION on A determined by the polynomial p. Also for fixed $a \in A$ define

$$\tau_a \colon \mathbb{F}[x] \to A \colon p \mapsto \widetilde{p}(a).$$

The mapping τ_a is the POLYNOMIAL FUNCTIONAL CALCULUS determined by the element a.

It is important to distinguish between the concepts of polynomials with coefficients in an algebra and polynomial functions. Also important is the distinction between the indeterminant x in $l(\mathbb{Z}^+, A)$ and x used as a variable for a polynomial function. (See 4.5.18.)

Exercise 4.5.15. Let A be a unital commutative algebra over a field $\mathbb{F}$. Then for each $a \in A$ the polynomial functional calculus $\tau_a \colon \mathbb{F}[x] \to A$ defined in 4.5.14 is a unital algebra homomorphism.

Proposition 4.5.16. *Let A be a unital commutative algebra over a field $\mathbb{F}$. The map*

$$\Psi \colon \mathbb{F}[x] \to l(A, A) \colon p \mapsto \widetilde{p}$$

is a unital algebra homomorphism.

Exercise 4.5.17. Under the homomorphism Ψ (defined in 4.5.16) what is the image of the indeterminant x? Under the homomorphism τ_a (defined in 4.5.14) what is the image of the indeterminant x?

The following example is intended to illustrate the importance of distinguishing between polynomials and polynomial functions.

Example 4.5.18. Let $\mathbb{F} = \{0, 1\}$ be the two-element field. The polynomials $p = x + x^2 + x^3$ and $q = x$ in the polynomial algebra $\mathbb{F}[x]$ show that the homomorphism Ψ (defined in 4.5.16) need not be injective.

Proposition 4.5.19. *Let A be a unital algebra of finite dimension m over a field $\mathbb{F}$. For every $a \in A$ there exists a polynomial $p \in \mathbb{F}[x]$ such that $1 \le \deg p \le m$ and $\widetilde{p}(a) = 0$.*

4.6. Minimal Polynomials

Definition 4.6.1. Let V be a vector space over a field $\mathbb{F}$ and $T \in \mathfrak{L}(V)$. A nonzero polynomial $p \in \mathbb{F}[x]$ such that $\widetilde{p}(T) = 0$ is an ANNIHILATING POLYNOMIAL for T. A monic polynomial of smallest degree that annihilates T is a MINIMAL POLYNOMIAL for T.

Proposition 4.6.2. *Let V be a finite dimensional vector space over a field $\mathbb{F}$. Then every $T \in \mathfrak{L}(V)$ has a minimal polynomial.*

Hint for proof. Use Example 4.5.19.

Theorem 4.6.3 (Division Algorithm). *Let f and d be polynomials with coefficients in a field $\mathbb{F}$ with $d \neq 0$. Then there exist unique polynomials q and r in $\mathbb{F}[x]$ such that*

(i) $f = dq + r$ *and*
(ii) $r = 0$ *or* $\deg r < \deg d$.

Hint for proof. Let $f = \sum_{j=0}^{k} f_j x^j$ and $d = \sum_{j=0}^{m} d_j x^j$ be in standard form. The case $k < m$ is trivial. For $k \ge m$ suppose the result to be true for all polynomials of degree strictly less than k. What can you say about $\hat{f} = f - p$ where $p = (f_k \, d_m^{-1}) \, x^{k-m} d$?

Notation 4.6.4. If T is an operator on a finite dimensional vector space over a field $\mathbb{F}$, we denote its minimal polynomial in $\mathbb{F}[x]$ by m_T.

Proposition 4.6.5. *Let V be a finite dimensional vector space over a field $\mathbb{F}$ and $T \in \mathfrak{L}(V)$. Then the minimal polynomial m_T for T is unique.*

Proposition 4.6.6. *An operator T on a finite dimensional vector space is invertible if and only if the constant term of its minimal polynomial is not zero.*

Exercise 4.6.7. Explain how, for an invertible operator T on a finite dimensional vector space, we can write its inverse as a polynomial in T.

Definition 4.6.8. If $\mathbb{F}$ is a field and p, $p_1 \in \mathbb{F}[x]$, we say that p_1 DIVIDES p if there exists $q \in \mathbb{F}[x]$ such that $p = p_1 q$.

Proposition 4.6.9. *Let T be an operator on a finite dimensional vector space over a field $\mathbb{F}$. If $p \in \mathbb{F}[x]$ and $\widetilde{p}(T) = 0$, then m_T divides p.*

Definition 4.6.10. A polynomial $p \in \mathbb{F}[x]$ is REDUCIBLE over $\mathbb{F}$ if there exist polynomials f, $g \in \mathbb{F}[x]$ both of degree at least one such that $p = fg$. A polynomial p of degree at least one is IRREDUCIBLE (or PRIME) over $\mathbb{F}$ provided that whenever $p = fg$ with f, $g \in \mathbb{F}[x]$, then either f or g is constant. That is, a polynomial p of degree at least one is irreducible if and only if it is not reducible.

Example 4.6.11. Let T be the operator on the real vector space $\mathbb{R}^2$ whose matrix representation (with respect to the standard basis) is $\begin{bmatrix} 0 & -1 \\ 1 & 0 \end{bmatrix}$. Find the minimal polynomial m_T of T and show that it is irreducible (over $\mathbb{R}$).

Example 4.6.12. Let T be the operator on the complex vector space $\mathbb{C}^2$ whose matrix representation (with respect to the standard basis) is $\begin{bmatrix} 0 & -1 \\ 1 & 0 \end{bmatrix}$. Find the minimal polynomial m_T of T and show that it is *reducible* (over $\mathbb{C}$).

Definition 4.6.13. A field $\mathbb{F}$ is ALGEBRAICALLY CLOSED if every prime polynomial in $\mathbb{F}[x]$ has degree 1.

Example 4.6.14. The field $\mathbb{R}$ of real numbers is not algebraically closed.

Proposition 4.6.15. *Let* $\mathbb{F}$ *be a field and* p *be a polynomial of degree* $m \geq 1$ *in* $\mathbb{F}[x]$. *If* J_p *is the principal ideal generated by* p *in* $\mathbb{F}[x]$, *then* $\dim \mathbb{F}[x]/J_p = m$.

Hint for proof. See Proposition 4.2.22. Show that $B = \{[x^k]: k = 0, 1, \ldots, m-1\}$ is a basis for the vector space $\mathbb{F}[x]/J_p$.

Corollary 4.6.16. *Let* T *be an operator on a finite dimensional vector space* V *over a field* $\mathbb{F}$, $\Phi\colon \mathbb{F}[x] \to \mathcal{L}(V)$ *be its associated polynomial functional calculus, and* J_{m_T} *be the principal ideal generated by its minimal polynomial. Then the sequence*

$$0 \longrightarrow J_{m_T} \longrightarrow \mathbb{F}[x] \overset{\Phi}{\longrightarrow} \operatorname{ran} \Phi \longrightarrow 0$$

is exact. Furthermore, the dimension of the range of the functional calculus associated with the operator T *is the degree of its minimal polynomial.*

Definition 4.6.17. Let $t_0, t_1, \ldots, t_n$ be distinct elements of a field $\mathbb{F}$. For $0 \leq k \leq n$ define $p_k \in \mathbb{F}[x]$ by

$$p_k = \prod_{\substack{j=0 \\ j \neq k}}^{n} \frac{x - t_j}{t_k - t_j}.$$

Proposition 4.6.18 (Lagrange Interpolation Formula). *The polynomials defined in 4.6.17 form a basis for the vector space* V *of all polynomials with coefficients in* $\mathbb{F}$ *and degree less than or equal to* n *and that for each polynomial* $q \in V$

$$q = \sum_{k=0}^{n} q(t_k)p_k.$$

Exercise 4.6.19. Use the *Lagrange Interpolation Formula* to find the polynomial with coefficients in $\mathbb{R}$ and degree no greater than 3 whose values at -1, 0, 1, and 2 are, respectively, -6, 2, -2, and 6.

Proposition 4.6.20. *Let* $\mathbb{F}$ *be a field and* p, q, *and* r *be polynomials in* $\mathbb{F}[x]$. *If* p *is a prime in* $\mathbb{F}[x]$ *and* p *divides* qr, *then* p *divides* q *or* p *divides* r.

Proposition 4.6.21. *Let* $\mathbb{F}$ *be a field. Then every nonzero ideal in* $\mathbb{F}[x]$ *is principal.*

Hint for proof. If J is a nonzero ideal in $\mathbb{F}[x]$ consider the principal ideal generated by any member of J of smallest degree.

Definition 4.6.22. Let $p_1, \ldots, p_n$ be polynomials, not all zero, with coefficients in a field $\mathbb{F}$. A monic polynomial d such that d divides each p_k ($k = 1, \ldots, n$) and such that any polynomial which divides each p_k also divides d is the GREATEST COMMON DIVISOR of the p_k's. The polynomials p_k are RELATIVELY PRIME if their greatest common divisor is 1.

Proposition 4.6.23. *Any finite set of polynomials (not all zero), with coefficients in a field $\mathbb{F}$ has a greatest common divisor.*

Theorem 4.6.24 (Unique Factorization). *Let $\mathbb{F}$ be a field. A nonconstant monic polynomial in $\mathbb{F}[x]$ can be factored in exactly one way (except for the order of the factors) as a product of monic primes in $\mathbb{F}[x]$.*

Definition 4.6.25. Let $\mathbb{F}$ be a field and $p(x) \in \mathbb{F}[x]$. An element $r \in \mathbb{F}$ is a ROOT of $p(x)$ if $p(r) = 0$.

Proposition 4.6.26. *Let $\mathbb{F}$ be a field and $p(x) \in \mathbb{F}[x]$. Then r is a root of $p(x)$ if and only if $x - r$ is a factor of $p(x)$.*

4.7. Invariant Subspaces

Definition 4.7.1. Let T be an operator on a vector space V. A subspace M of V is INVARIANT UNDER T (or T-INVARIANT) if $T^{\rightarrow}(M) \subseteq M$. Since the subspaces $\{0\}$ and V are invariant under any operator on V, they are called the TRIVIAL invariant subspaces.

Exercise 4.7.2. Let S be the operator on $\mathbb{R}^3$ whose matrix representation is $\begin{bmatrix} 3 & 4 & 2 \\ 0 & 1 & 2 \\ 0 & 0 & 0 \end{bmatrix}$. Find three one dimensional subspaces U, V, and W of $\mathbb{R}^3$ that are invariant under S.

Exercise 4.7.3. Let T be the operator on $\mathbb{R}^3$ whose matrix representation is $\begin{bmatrix} 0 & 0 & 2 \\ 0 & 2 & 0 \\ 2 & 0 & 0 \end{bmatrix}$. Find a two dimensional subspace U of $\mathbb{R}^3$ that is invariant under T.

Exercise 4.7.4. Find infinitely many subspaces of the vector space of polynomial functions on $\mathbb{R}$ that are invariant under the differentiation operator.

Definition 4.7.5. An operator T on a vector space V is REDUCED by a pair (M, N) of subspaces M and N of V if $V = M \oplus N$ and both M and N are invariant under T. In this case M and N are REDUCING SUBSPACES for T.

Exercise 4.7.6. Let T be the operator on $\mathbb{R}^3$ whose matrix representation is $\begin{bmatrix} 2 & 0 & 0 \\ -1 & 3 & 2 \\ 1 & -1 & 0 \end{bmatrix}$. Find a plane and a line in $\mathbb{R}^3$ that reduce T.

Proposition 4.7.7. *Let M be a subspace of a vector space V and $T \in \mathcal{L}(V)$. If M is invariant under T, then $ETE = TE$ for every projection E onto M. And if $ETE = TE$ for some projection E onto M, then M is invariant under T.*

Proposition 4.7.8. *Suppose a vector space V has the direct sum decomposition $V = M \oplus N$. Then an operator T on V is reduced by the pair (M, N) if and only if $ET = TE$, where $E = E_{MN}$ is the projection along M onto N.*

Proposition 4.7.9. *Suppose a finite dimensional vector space V has the direct sum decomposition $V = M \oplus N$ and that $E = E_{MN}$ is the projection along M onto N. Show that E^* is the projection in $\mathcal{L}(V^*)$ along $N^\perp$ onto $M^\perp$.*

Proposition 4.7.10. *Let M and N be complementary subspaces of a vector space V (that is, V is the direct sum of M and N) and let T be an operator on V. If M is invariant under T, then $M^\perp$ is invariant under T^* and if T is reduced by the pair (M, N), then T^* is reduced by the pair $(M^\perp, N^\perp)$.*

4.8. Burnside's Theorem

Notation 4.8.1. Let V be a vector space. For $T \in \mathcal{L}(V)$ let

$$\operatorname{Lat} T := \{M \preceq V : M \text{ is invariant under } T\}.$$

If $\mathfrak{T} \subseteq \mathfrak{L}(V)$ let

$$\operatorname{Lat} \mathfrak{T} := \bigcap_{T \in \mathfrak{T}} \operatorname{Lat} T.$$

We say that $\operatorname{Lat} T$ (or $\operatorname{Lat} \mathfrak{T}$) is TRIVIAL if it contains only the trivial invariant subspaces $\{0\}$ and V.

Example 4.8.2. If V is a vector space, then $\operatorname{Lat} \mathfrak{L}(V)$ is trivial.

Hint for proof. For $\dim V \geq 2$ let M be a nonzero proper subspace of V. Choose nonzero vectors $x \in M$ and $y \in M^c$. Define $T: V \to V: v \mapsto f(v)y$ where f is a functional in V^* such that $f(x) = 1$.

Example 4.8.3. Let $\mathfrak{A}$ be the subalgebra of $\mathfrak{L}(\mathbb{R}^2)$ whose members have matrix representations of the form $\begin{bmatrix} a & b \\ -b & a \end{bmatrix}$. Then $\operatorname{Lat} \mathfrak{A}$ is trivial.

Example 4.8.4. Let $\mathfrak{A}$ be the subalgebra of $\mathfrak{L}(\mathbb{C}^2)$ whose members have matrix representations of the form $\begin{bmatrix} a & b \\ -b & a \end{bmatrix}$. Then $\operatorname{Lat} \mathfrak{A}$ is not trivial.

Hint for proof. Try $\operatorname{span}\{(1, -i)\}$.

Definition 4.8.5. Let V be a vector space. A subalgebra $\mathfrak{A}$ of $\mathfrak{L}(V)$ is TRANSITIVE if for every $x \neq 0$ and y in V there exists an operator T in $\mathfrak{A}$ such that $y = Tx$.

Proposition 4.8.6. *Let V be a vector space. A subalgebra $\mathfrak{A}$ of $\mathfrak{L}(V)$ is transitive if and only if $\operatorname{Lat} \mathfrak{A}$ is trivial.*

Definition 4.8.7. A field $\mathbb{F}$ is ALGEBRAICALLY CLOSED if every nonconstant polynomial in $\mathbb{F}[x]$ has a root in $\mathbb{F}$.

Example 4.8.8. The field $\mathbb{C}$ of complex numbers is algebraically closed; the field $\mathbb{R}$ of real numbers is not.

Theorem 4.8.9 (Burnside's Theorem). *Let V be a finite dimensional vector space over an algebraically closed field. Then $\mathfrak{L}(V)$ has no proper subalgebra that is transitive.*

Proof. See [13], Theorem 3.15.

Corollary 4.8.10. *Let V be a finite dimensional complex vector space of dimension at least 2. Then every proper subalgebra of $\mathfrak{L}(V)$ has a nontrivial invariant subspace.*

Example 4.8.11. The preceding result does not hold for real vector spaces.

4.9. Eigenvalues and Eigenvectors

Definition 4.9.1. Suppose that on a vector space V there exist projection operators $E_1, \ldots, E_n$ such that

(i) $I_V = E_1 + E_2 + \cdots + E_n$ and
(ii) $E_i E_j = 0$ whenever $i \neq j$.

Then we say that the family $\{E_1, E_2, \ldots, E_n\}$ of projections is a RESOLUTION OF THE IDENTITY.

Recall that it was shown in Proposition 4.1.4 that if $\{E_1, E_2, \ldots, E_n\}$ is a resolution of the identity on a vector space V, then $V = \bigoplus_{k=1}^{n} \operatorname{ran} E_k$.

Definition 4.9.2. Let $M_1 \oplus \cdots \oplus M_n$ be a direct sum decomposition of a vector space V. For each $k \in \mathbb{N}_n$ let N_k be the following subspace of V complementary to M_k:

$$N_k := M_1 \oplus \cdots \oplus M_{k-1} \oplus M_{k+1} \oplus \cdots \oplus M_n.$$

Also (for each k) let

$$E_k := E_{N_k M_k}$$

be the projection onto M_k along the complementary subspace N_k. The projections $E_1, \ldots, E_n$ are the PROJECTIONS ASSOCIATED WITH THE DIRECT SUM DECOMPOSITION $V = M_1 \oplus \cdots \oplus M_n$.

Proposition 4.9.3. *If $M_1 \oplus \cdots \oplus M_n$ is a direct sum decomposition of a vector space V, then the family $\{E_1, E_2, \ldots, E_n\}$ of the associated projections is a resolution of the identity.*

In the following definition we make use of the familiar notion of the *determinant* of a matrix even though we have not yet developed the theory of determinants. We will eventually do this.

Definition 4.9.4. Let V be a vector space over a field $\mathbb{F}$ and $T \in \mathfrak{L}(V)$. An element $\lambda \in \mathbb{F}$ is an EIGENVALUE of T if $\ker(T - \lambda I) \neq \{\mathbf{0}\}$. The collection of all eigenvalues of T is its POINT SPECTRUM, denoted by $\sigma_p(T)$.

Definition 4.9.5. If $\mathbb{F}$ is a field and A is an $n \times n$ matrix of elements of $\mathbb{F}$, we define the CHARACTERISTIC POLYNOMIAL c_A of A to be the determinant of $A - xI$. (Note that $\det(A - xI)$ is a polynomial.) Some authors prefer the characteristic polynomial to be monic, and consequently define it to be the determinant of $xI - A$. As you would expect, the characteristic polynomial c_T of an operator T on a finite dimensional space (with basis B) is the characteristic polynomial of the matrix representation of that operator (with respect to B). Making use of some standard facts (which we have not yet proved) about determinants (see Section 7.3) we see that $\lambda \in \mathbb{F}$ is an eigenvalue of the matrix A (or of its associated linear transformation) if and only if it is a root of the characteristic polynomial c_A.

Proposition 4.9.6. *If T is an operator on a finite dimensional vector space, then $\sigma_p(T) = \sigma(T)$.*

Exercise 4.9.7. Let $A = \begin{bmatrix} 1 & 1 & 1 \\ 1 & 1 & 1 \\ 1 & 1 & 1 \end{bmatrix}$.

The characteristic polynomial of A is $\lambda^p(\lambda - 3)^q$ where $p =$ _____ and $q =$ _____ .

The minimal polynomial of A is $\lambda^r(\lambda - 3)^s$ where $r =$ _____ and $s =$ _____ .

Exercise 4.9.8. Let T be the operator on $\mathbb{R}^4$ whose matrix representation is $\begin{bmatrix} 0 & 1 & 0 & -1 \\ -2 & 3 & 0 & -1 \\ -2 & 1 & 2 & -1 \\ 2 & -1 & 0 & 3 \end{bmatrix}$.

The characteristic polynomial of T is $(\lambda - 2)^p$ where $p =$ _____ .

The minimal polynomial of T is $(\lambda - 2)^r$ where $r =$ _____ .

Exercise 4.9.9. Choose a, b and c in the matrix $A = \begin{bmatrix} 0 & 1 & 0 \\ 0 & 0 & 1 \\ a & b & c \end{bmatrix}$ so that the characteristic polynomial of A is $-\lambda^3 + 4\lambda^2 + 5\lambda + 6$.

Proposition 4.9.10. *Let V be a finite dimensional vector space over a field $\mathbb{F}$. An operator T on V is invertible if and only if T is not a zero divisor in $\mathfrak{L}(V)$.*

Hint for proof. In one direction the proof is very easy. For the other, use Proposition 4.6.6.

Corollary 4.9.11. *Let V be a finite dimensional vector space over a field $\mathbb{F}$, $T \in \mathfrak{L}(V)$, and $\lambda \in \mathbb{F}$. Then $T - \lambda I$ fails to be invertible if and only if λ is a root of the minimal polynomial of T.*

Corollary 4.9.12. *If T is an operator on a finite dimensional vector space, then its minimal polynomial and characteristic polynomial have the same roots.*

Theorem 4.9.13 (Cayley-Hamilton Theorem). *If T is an operator on a finite dimensional vector space, then the characteristic polynomial of T annihilates T. Moreover, the minimal polynomial of T is a factor of the characteristic polynomial of T.*

Proof. See [13], Proposition 3.19.

Definition 4.9.14. Let V be a vector space, T be an operator on V, and λ be an eigenvalue of T. A nonzero vector x in the kernel of $T - \lambda I$ is an EIGENVECTOR OF T ASSOCIATED WITH (or CORRESPONDING TO, or BELONGING TO) the eigenvalue λ.

Definition 4.9.15. Let V be a vector space, T be an operator on V, and λ be an eigenvalue of T. The EIGENSPACE ASSOCIATED WITH (or CORRESPONDING TO, or BELONGING TO) the eigenvalue λ is the kernel of $T - \lambda I$.

Exercise 4.9.16. Let T be the operator on $\mathbb{R}^3$ whose matrix representation is $\begin{bmatrix} 3 & 1 & -1 \\ 2 & 2 & -1 \\ 2 & 2 & 0 \end{bmatrix}$.

(a) Find the characteristic polynomial of T.
(b) Find the minimal polynomial of T.
(c) Find the eigenspaces V_1 and V_2 of T.

Exercise 4.9.17. Let T be the operator on $\mathbb{R}^5$ whose matrix representation is $\begin{bmatrix} 1 & 0 & 0 & 1 & -1 \\ 0 & 1 & -2 & 3 & -3 \\ 0 & 0 & -1 & 2 & -2 \\ 1 & -1 & 1 & 0 & 1 \\ 1 & -1 & 1 & -1 & 2 \end{bmatrix}$.

(a) Find the characteristic polynomial of T.

(b) Find the minimal polynomial of T.

Proposition 4.9.18. *If $\lambda_1 \neq \lambda_2$ are eigenvalues of an operator T, then the eigenspaces M_1 and M_2 associated with λ_1 and λ_2, respectively, have only 0 in common.*

Proposition 4.9.19. *Let V be a vector space over a field $\mathbb{F}$. If v is an eigenvector associated with an eigenvalue λ of an operator $T \in \mathfrak{L}(V)$ and p is a polynomial in $\mathbb{F}[x]$, then $p(T)v = p(\lambda)v$.*

Definition 4.9.20. Two operators on a vector space (or two $n \times n$ matrices) R and T are SIMILAR if there exists an invertible operator (or matrix) S such that $R = S^{-1}TS$.

Proposition 4.9.21. *If R and T are operators on a vector space and R is invertible, then RT is similar to TR.*

Example 4.9.22. If R and T are operators on a vector space, then RT need not be similar to TR.

Proposition 4.9.23. *Let R and T be operators on a vector space. If R is similar to T and $p \in \mathbb{F}[x]$ is a polynomial, then $p(R)$ is similar to $p(T)$.*

Proposition 4.9.24. *If R and T are operators on a vector space, R is similar to T, and R is invertible, then T is invertible and T^{-1} is similar to R^{-1}.*

Proposition 4.9.25. *If two matrices A and B are similar, then they have the same spectrum.*

Hint for proof. You may use familiar facts about determinants that we have not yet proved.

Definition 4.9.26. An operator on a vector space is NILPOTENT if some power of the operator is 0.

Proposition 4.9.27. *An operator T on a finite dimensional complex vector space is nilpotent if and only if $\sigma(T) = \{0\}$.*

Notation 4.9.28. Let $\alpha_1, \ldots, \alpha_n$ be elements of a field $\mathbb{F}$. Then $\operatorname{diag}(\alpha_1, \ldots, \alpha_n)$ denotes the $n \times n$ matrix whose entries are all zero except on the main diagonal where they are $\alpha_1, \ldots, \alpha_n$. Such a matrix is a DIAGONAL matrix.

Definition 4.9.29. Let V be a vector space of finite dimension n. An operator T on V is DIAGONALIZABLE if it has n linearly independent eigenvectors (or, equivalently, if V has a basis of eigenvectors of T).

Proposition 4.9.30. *Let A be an $n \times n$ matrix with entries from a field $\mathbb{F}$. Then A, regarded as an operator on $\mathbb{F}^n$, is diagonalizable if and only if it is similar to a diagonal matrix.*

4.10. The Spectral Theorem — Vector Space Version

Proposition 4.10.1. *Let $E_1, \ldots, E_n$ be the projections associated with a direct sum decomposition $V = M_1 \oplus \cdots \oplus M_n$ of a vector space V and let T be an operator on V. Then each subspace M_k is invariant under T if and only if T commutes with each projection E_k.*

Theorem 4.10.2 (Spectral Theorem for Vector Spaces). *If T is a diagonalizable operator on a finite dimensional vector space V, then*

$$T = \sum_{k=1}^{n} \lambda_k E_k$$

where $\lambda_1, \ldots, \lambda_n$ are the (distinct) eigenvalues of T and $\{E_1, \ldots, E_n\}$ is the resolution of the identity whose projections are associated with the corresponding eigenspaces $M_1, \ldots, M_n$.

Proposition 4.10.3. *Let T be an operator on a finite dimensional vector space V. If $\lambda_1, \ldots, \lambda_n$ are distinct scalars and $E_1, \ldots, E_n$ are nonzero operators on V such that*

(i) $T = \sum_{k=1}^{n} \lambda_k E_k$,
(ii) $I = \sum_{k=1}^{n} E_k$, *and*
(iii) $E_j E_k = \mathbf{0}$ *whenever $j \neq k$,*

then T is diagonalizable, the scalars $\lambda_1, \ldots, \lambda_n$ are the eigenvalues of T, and the operators $E_1, \ldots, E_n$ are projections whose ranges are the eigenspaces of T.

Proposition 4.10.4. *If T is a diagonalizable operator on a finite dimensional vector space V over a field $\mathbb{F}$ and $p \in \mathbb{F}[x]$, then*

$$p(T) = \sum_{k=1}^{n} p(\lambda_k) E_k$$

where $\lambda_1, \ldots, \lambda_n$ are the (distinct) eigenvalues of T and $E_1, \ldots, E_n$ are the projections associated with the corresponding eigenspaces $M_1, \ldots, M_n$.

Proposition 4.10.5. *If T is a diagonalizable operator on a finite dimensional vector space V, then the projections $E_1, \ldots, E_n$ associated with the decomposition of V as a direct sum $\bigoplus M_k$ of its eigenspaces can be expressed as polynomials in T.*

Hint for proof. Apply the *Lagrange interpolation formula (Proposition 4.6.18)* with the t_k's being the eigenvalues of T.

Exercise 4.10.6. Let T be the operator on $\mathbb{R}^3$ whose matrix representation is $\begin{bmatrix} 0 & 0 & 2 \\ 0 & 2 & 0 \\ 2 & 0 & 0 \end{bmatrix}$. Use Proposition 4.10.5 to write T as a linear combination of projections.

Exercise 4.10.7. Let T be the operator on $\mathbb{R}^3$ whose matrix representation is $\begin{bmatrix} 2 & -2 & 1 \\ -1 & 1 & 1 \\ -1 & 2 & 0 \end{bmatrix}$. Use Proposition 4.10.5 to write T as a linear combination of projections.

Exercise 4.10.8. Let T be the operator on $\mathbb{R}^3$ whose matrix representation is

$$\begin{bmatrix} \frac{1}{3} & -\frac{2}{3} & -\frac{2}{3} \\ -\frac{2}{3} & \frac{5}{6} & -\frac{7}{6} \\ -\frac{2}{3} & -\frac{7}{6} & \frac{5}{6} \end{bmatrix}.$$

Write T as a linear combination of projections.

Proposition 4.10.9. *An operator T on a finite dimensional vector space is diagonalizable if and only if its minimal polynomial is of the form $\prod_{k=1}^{n}(x - \lambda_k)$ for some distinct elements $\lambda_1, \ldots, \lambda_n$ of the scalar field $\mathbb{F}$.*

Proof. See [16], page 204, Theorem 6.

4.11. Two Decomposition Theorems

Theorem 4.11.1 (Primary Decomposition Theorem). *Let $T \in \mathfrak{L}(V)$ where V is a finite dimensional vector space. Factor the minimal polynomial*

$$m_T = \prod_{k=1}^{n} p_k^{r_k}$$

into powers of distinct irreducible monic polynomials $p_1, \ldots, p_n$ and let $W_k = \ker\left(p_k(T)\right)^{r_k}$ for each k. Then

(i) $V = \bigoplus_{k=1}^{n} W_k$,

(ii) each W_k is invariant under T, and

(iii) if $T_k = T\big|_{W_k}$, then $m_{T_k} = p_k^{r_k}$.

Proof. See [16], page 220, Theorem 12.

In the preceding theorem the spaces W_k are the GENERALIZED EIGENSPACES of the operator T.

Theorem 4.11.2 (Diagonalizable Plus Nilpotent Decomposition).
Let T be an operator on a finite dimensional vector space V. Suppose that the minimal polynomial for T factors completely into linear factors

$$m_T(x) = (x - \lambda_1)^{d_1} \cdots (x - \lambda_r)^{d_r}$$

where $\lambda_1, \ldots, \lambda_r$ are the (distinct) eigenvalues of T. For each k let W_k be the generalized eigenspace $\ker(T - \lambda_k I)^{d_k}$ and let $E_1, \ldots, E_r$ be the projections associated with the direct sum decomposition

$$V = W_1 \oplus W_2 \oplus \cdots \oplus W_r.$$

Then this family of projections is a resolution of the identity, each W_k is invariant under T, the operator

$$D = \lambda_1 E_1 + \cdots + \lambda_r E_r$$

is diagonalizable, the operator

$$N = T - D$$

is nilpotent, and N commutes with D.

Furthermore, if D_1 is diagonalizable, N_1 is nilpotent, $D_1 + N_1 = T$, and $D_1 N_1 = N_1 D_1$, then $D_1 = D$ and $N_1 = N$.

Proof. See [16], page 222, Theorem 13.

Corollary 4.11.3. *Every operator on a finite dimensional complex vector space can be written as the sum of two commuting operators, one diagonalizable and the other nilpotent.*

Exercise 4.11.4. Let T be the operator on $\mathbb{R}^2$ whose matrix representation is $\begin{bmatrix} 2 & 1 \\ -1 & 4 \end{bmatrix}$.

(a) Explain briefly why T is not diagonalizable.

(b) Find the diagonalizable and nilpotent parts of T.

Answer: $D = \begin{bmatrix} a & b \\ b & a \end{bmatrix}$ and $N = \begin{bmatrix} -c & c \\ -c & c \end{bmatrix}$ where $a = \underline{\quad}$, $b = \underline{\quad}$,

and $c = \underline{\quad}$.

Exercise 4.11.5. Let T be the operator on $\mathbb{R}^3$ whose matrix representation is $\begin{bmatrix} 0 & 0 & -3 \\ -2 & 1 & -2 \\ 2 & -1 & 5 \end{bmatrix}$.

(a) Find D and N, the diagonalizable and nilpotent parts of T. Express these as polynomials in T.

(b) Find a matrix S which diagonalizes D.

(c) Let $[D_1] = \begin{bmatrix} 2 & -1 & -1 \\ -1 & 2 & -1 \\ -1 & -1 & 2 \end{bmatrix}$ and $[N_1] = \begin{bmatrix} -2 & 1 & -2 \\ -1 & -1 & -1 \\ 3 & 0 & 3 \end{bmatrix}$. Show that D_1 is diagonalizable, that N_1 is nilpotent, and that $T = D_1 + N_1$. Why does this not contradict the uniqueness claim made in Theorem 4.11.2?

Exercise 4.11.6. Let T be the operator on $\mathbb{R}^4$ whose matrix representation is $\begin{bmatrix} 0 & 1 & 0 & -1 \\ -2 & 3 & 0 & -1 \\ -2 & 1 & 2 & -1 \\ 2 & -1 & 0 & 3 \end{bmatrix}$.

(a) The characteristic polynomial of T is $(\lambda - 2)^p$ where $p = \underline{\quad}$.

(b) The minimal polynomial of T is $(\lambda - 2)^r$ where $r = \underline{\quad}$.

(c) The diagonalizable part of T is $D = \begin{bmatrix} a & b & b & b \\ b & a & b & b \\ b & b & a & b \\ b & b & b & a \end{bmatrix}$ where $a = \underline{\quad}$ and $b = \underline{\quad}$.

(d) The nilpotent part of T is $N = \begin{bmatrix} -a & b & c & -b \\ -a & b & c & -b \\ -a & b & c & -b \\ a & -b & c & b \end{bmatrix}$ where $a = \underline{\quad}$, $b = \underline{\quad}$, and $c = \underline{\quad}$.

Exercise 4.11.7. Let T be the operator on $\mathbb{R}^5$ whose matrix representation is
$$\begin{bmatrix} 1 & 0 & 0 & 1 & -1 \\ 0 & 1 & -2 & 3 & -3 \\ 0 & 0 & -1 & 2 & -2 \\ 1 & -1 & 1 & 0 & 1 \\ 1 & -1 & 1 & -1 & 2 \end{bmatrix}.$$

(a) Find the characteristic polynomial of T.

Answer: $c_T(\lambda) = (\lambda + 1)^p (\lambda - 1)^q$ where $p =$ ___ and $q =$ ___ .

(b) Find the minimal polynomial of T.

Answer: $m_T(\lambda) = (\lambda + 1)^r (\lambda - 1)^s$ where $r =$ ___ and $s =$ ___ .

(c) Find the eigenspaces V_1 and V_2 of T.

Answer: $V_1 = \text{span}\{(a, 1, b, a, a)\}$ where $a =$ ___ and $b =$ ___ ; and

$V_2 = \text{span}\{(1, a, b, b, b), (b, b, b, 1, a)\}$ where $a =$ ___ and $b =$ ___ .

(d) Find the diagonalizable part of T.

Answer: $D = \begin{bmatrix} a & b & b & b & b \\ b & a & -c & c & -c \\ b & b & -a & c & -c \\ b & b & b & a & b \\ b & b & b & b & a \end{bmatrix}$ where $a =$ ___ , $b =$ ___ , and $c =$ ___ .

(e) Find the nilpotent part of T.

Answer: $N = \begin{bmatrix} a & a & a & b & -b \\ a & a & a & b & -b \\ a & a & a & a & a \\ b & -b & b & -b & b \\ b & -b & b & -b & b \end{bmatrix}$ where $a =$ ___ and $b =$ ___ .

(f) Find a matrix S which diagonalizes the diagonalizable part D of T. What is the diagonal form Λ of D associated with this matrix?

Answer: $S = \begin{bmatrix} a & b & a & a & a \\ b & a & b & a & a \\ b & a & a & b & a \\ a & a & a & b & b \\ a & a & a & a & b \end{bmatrix}$ where $a =$ ___ and $b =$ ___ .

and $\Lambda = \begin{bmatrix} -a & 0 & 0 & 0 & 0 \\ 0 & a & 0 & 0 & 0 \\ 0 & 0 & a & 0 & 0 \\ 0 & 0 & 0 & a & 0 \\ 0 & 0 & 0 & 0 & a \end{bmatrix}$ where $a =$ ___ .

Chapter 5

THE SPECTRAL THEOREM
FOR INNER PRODUCT SPACES

In this chapter all vector spaces (and algebras) have complex or
real scalars.

5.1. Inner Products

Definition 5.1.1. Let V be a complex (or a real) vector space. A function
that associates to each pair of vectors x and y in V a complex number (or, in
the case of a real vector space, a real number) $\langle x, y \rangle$ is an INNER PRODUCT
(or a DOT PRODUCT) on V provided that the following four conditions are
satisfied:

(a) If x, y, $z \in V$, then $\langle x + y, z \rangle = \langle x, z \rangle + \langle y, z \rangle$.
(b) If x, $y \in V$, then $\langle \alpha x, y \rangle = \alpha \langle x, y \rangle$.
(c) If x, $y \in V$, then $\langle x, y \rangle = \overline{\langle y, x \rangle}$.
(d) For every nonzero x in V we have $\langle x, x \rangle > 0$.

Conditions (a) and (b) show that an inner product is linear in its first variable. Conditions (a) and (b) of Proposition 5.1.2 say that an inner product is
CONJUGATE LINEAR in its second variable. When a mapping is linear in one
variable and conjugate linear in the other, it is often called SESQUILINEAR
(the prefix "sesqui-" means "one and a half"). Taken together conditions
(a)–(d) say that the inner product is a *positive definite conjugate symmetric
sesquilinear form*. Of course, in the case of a real vector space, the complex

conjugation indicated in (c) has no effect and the inner product is a *positive definite symmetric bilinear form*. A vector space on which an inner product has been defined is an INNER PRODUCT SPACE.

Proposition 5.1.2. *If x, y, and z are vectors in an inner product space and $\alpha \in \mathbb{C}$, then*

(a) $\langle x, y + z \rangle = \langle x, y \rangle + \langle x, z \rangle$,
(b) $\langle x, \alpha y \rangle = \overline{\alpha} \langle x, y \rangle$, *and*
(c) $\langle x, x \rangle = 0$ *if and only if $x = 0$.*

Example 5.1.3. For vectors $x = (x_1, x_2, \ldots, x_n)$ and $y = (y_1, y_2, \ldots, y_n)$ belonging to $\mathbb{C}^n$ define

$$\langle x, y \rangle = \sum_{k=1}^{n} x_k \overline{y_k} \,.$$

Then $\mathbb{C}^n$ is an inner product space.

Example 5.1.4. For $a < b$ let $\mathcal{C}([a, b], \mathbb{C})$ be the family of all continuous complex valued functions on the interval $[a, b]$. For every f, $g \in \mathcal{C}([a, b], \mathbb{C})$ define

$$\langle f, g \rangle = \int_a^b f(x)\overline{g(x)} \, dx.$$

Then $\mathcal{C}([a, b], \mathbb{C})$ is a complex inner product space. In a similar fashion, $\mathcal{C}([a, b]) = \mathcal{C}([a, b], \mathbb{R})$ is made into a real inner product space.

Definition 5.1.5. When a (real or complex) vector space has been equipped with an inner product we define the NORM of a vector x by

$$\|x\| := \sqrt{\langle x, x \rangle};$$

(This somewhat optimistic terminology is justified in Proposition 5.1.13 below.)

Theorem 5.1.6. *In every inner product space the* Schwarz inequality

$$|\langle x, y \rangle| \leq \|x\| \, \|y\|.$$

holds for all vectors x and y.

Hint for proof. Let V be an inner product space and fix vectors x, $y \in V$. For every scalar α we know that

$$0 \le \langle x - \alpha y, x - \alpha y \rangle. \tag{5.1}$$

Expand the right hand side of (5.1) into four terms and write $\langle y, x \rangle$ in polar form: $\langle y, x \rangle = re^{i\theta}$, where $r > 0$ and $\theta \in \mathbb{R}$. Then in the resulting inequality consider those α of the form $te^{-i\theta}$ where $t \in \mathbb{R}$. Notice that now the right side of (5.1) is a quadratic polynomial in t. What can you say about its discriminant?

Exercise 5.1.7. If $a_1, \dots, a_n > 0$, then

$$\left(\sum_{j=1}^{n} a_j \right) \left(\sum_{k=1}^{n} \frac{1}{a_k} \right) \ge n^2.$$

The proof of this is obvious from the Schwarz inequality if we choose x and y to be what?

Exercise 5.1.8. In this exercise notice that part (a) is a special case of part (b).

(a) Show that if a, b, $c > 0$, then $\left(\frac{1}{2}a + \frac{1}{3}b + \frac{1}{6}c \right)^2 \le \frac{1}{2}a^2 + \frac{1}{3}b^2 + \frac{1}{6}c^2$.

(b) Show that if $a_1, \dots, a_n, w_1, \dots, w_n > 0$ and $\sum_{k=1}^{n} w_k = 1$, then

$$\left(\sum_{k=1}^{n} a_k w_k \right)^2 \le \sum_{k=1}^{n} a_k^2 w_k.$$

Exercise 5.1.9. Show that if $\sum_{k=1}^{\infty} a_k^2$ converges, then $\sum_{k=1}^{\infty} k^{-1} a_k$ converges absolutely.

Example 5.1.10. A sequence (a_k) of (real or) complex numbers is said to be SQUARE SUMMABLE if $\sum_{k=1}^{\infty} |a_k|^2 < \infty$. The vector space of all square summable sequences of real numbers (respectively, complex numbers) is denoted by $l_2(\mathbb{R})$ (respectively, $l_2(\mathbb{C})$). When no confusion will result, both are denoted by l_2. If a, $b \in l^2$, define

$$\langle a, b \rangle = \sum_{k=1}^{\infty} a_k \overline{b_k}.$$

(It must be shown that this definition makes sense and that it makes l_2 into an inner product space.)

Definition 5.1.11. Let V be a complex (or real) vector space. A function $\| \ \|\colon V \to \mathbb{R}\colon x \mapsto \|x\|$ is a NORM on V if

(i) $\|x + y\| \leq \|x\| + \|y\|$ for all x, $y \in V$;
(ii) $\|\alpha x\| = |\alpha| \, \|x\|$ for all $x \in V$ and $\alpha \in \mathbb{C}$ (or $\mathbb{R}$); and
(iii) if $\|x\| = 0$, then $x = \mathbf{0}$.

The expression $\|x\|$ may be read as "the *norm* of x" or "the *length* of x".

A vector space on which a norm has been defined is a NORMED LINEAR SPACE (or NORMED VECTOR SPACE). A vector in a normed linear space that has norm 1 is a UNIT VECTOR.

Proposition 5.1.12. *If* $\| \ \|$ *is norm on a vector space* V, *then* $\|x\| \geq 0$ *for every* $x \in V$ *and* $\|\mathbf{0}\| = 0$.

As promised in Definition 5.1.5 we can verify the (somewhat obvious) fact that every inner product space is a normed linear space (and therefore a topological — in fact, a metric — space).

Proposition 5.1.13. *Let* V *be an inner product space. The map* $x \mapsto \|x\|$ *defined on* V *in 5.1.5 is a norm on* V.

Proposition 5.1.14 (The parallelogram law). *If* x *and* y *are vectors in an inner product space, then*

$$\|x + y\|^2 + \|x - y\|^2 = 2\|x\|^2 + 2\|y\|^2 \,.$$

Example 5.1.15. Consider the space $\mathcal{C}([0,1])$ of continuous complex valued functions defined on $[0,1]$. Under the UNIFORM NORM $\|f\|_u :=$ $\sup\{|f(x)|\colon 0 \leq x \leq 1\}$ the vector space $\mathcal{C}([0,1])$ is a normed linear space, but there is no inner product on $\mathcal{C}([0,1])$ that induces this norm.

Hint for proof. Use the preceding proposition.

Proposition 5.1.16 (The polarization identity). *If* x *and* y *are vectors in a complex inner product space, then*

$$\langle x, y \rangle = \tfrac{1}{4}\big(\|x + y\|^2 - \|x - y\|^2 + i\,\|x + iy\|^2 - i\,\|x - iy\|^2\big)\,.$$

Exercise 5.1.17. What is the corresponding formula for the *polarization identity* in a real inner product space?

5.2. Orthogonality

Definition 5.2.1. Vectors x and y in an inner product space H are ORTHOGONAL (or PERPENDICULAR) if $\langle x, y \rangle = 0$. In this case we write $x \perp y$. Subsets A and B of H are ORTHOGONAL if $a \perp b$ for every $a \in A$ and $b \in B$. In this case we write $A \perp B$.

Proposition 5.2.2. *Let a be a vector in an inner product space H. Then $a \perp x$ for every $x \in H$ if and only if $a = 0$.*

Proposition 5.2.3 (The Pythagorean theorem). *If $x \perp y$ in an inner product space, then*

$$\|x + y\|^2 = \|x\|^2 + \|y\|^2.$$

Definition 5.2.4. If M and N are subspaces of an inner product space H we use the notation $H = M \oplus N$ to indicate not only that H is the sum of M and N but also that M and N are orthogonal. Thus we say that H is the (INTERNAL) ORTHOGONAL DIRECT SUM of M and N.

Proposition 5.2.5. *If M and N are subspaces of an inner product space H and H is the orthogonal direct sum of M and N, then it is also the vector space direct sum of M and N.*

As is the case with vector spaces in general, we distinguish between internal and external direct sums.

Definition 5.2.6. Let V and W be inner product spaces. For (v, w) and (v', w') in $V \times W$ and $\alpha \in \mathbb{C}$ define

$$(v, w) + (v', w') = (v + v', w + w')$$

and

$$\alpha(v, w) = (\alpha v, \alpha w).$$

This results in a vector space, which is the *(external) direct sum* of V and W. To make it into an inner product space define

$$\langle (v, w), (v', w') \rangle = \langle v, v' \rangle + \langle w, w' \rangle.$$

This makes the direct sum of V and W into an inner product space. It is the (EXTERNAL ORTHOGONAL) DIRECT SUM of V and W and is denoted by $V \oplus W$.

Caution 5.2.7. Notice that the same notation $\oplus$ is used for both internal and external direct sums *and* for both vector space direct sums (see Definitions 1.5.14 and 3.4.3) and orthogonal direct sums. So when we see the symbol $V \oplus W$ it is important to be alert to context, to know which category we are in: vector spaces or inner product spaces, especially as it is common practice to omit the word "orthogonal" as a modifier to "direct sum" even in cases when it is intended.

Example 5.2.8. In $\mathbb{R}^2$ let M be the x-axis and L be the line whose equation is $y = x$. If we think of $\mathbb{R}^2$ as a (real) vector space, then it is correct to write $\mathbb{R}^2 = M \oplus L$. If, on the other hand, we regard $\mathbb{R}^2$ as a (real) inner product space, then $\mathbb{R}^2 \neq M \oplus L$ (because M and L are not perpendicular).

Notation 5.2.9. Let V be an inner product space, $x \in V$, and $A, B \subseteq V$. If $x \perp a$ for every $a \in A$, we write $x \perp A$; and if $a \perp b$ for every $a \in A$ and $b \in B$, we write $A \perp B$. We define $A^\perp$, the ORTHOGONAL COMPLEMENT of A, to be $\{x \in V : x \perp A\}$. We write $A^{\perp\perp}$ for $\left(A^\perp\right)^\perp$.

Caution 5.2.10. The superscript $\perp$ is here used quite differently than in our study of vector spaces (see 2.6.1). These two uses are, however, by no means unrelated! It is an instructive exercise to make explicit exactly what this relationship is.

Proposition 5.2.11. *If A is a subset of an inner product space V, then $A^\perp$ is a subspace of V and $A^\perp = (\text{span } A)^\perp$. Furthermore, if $A \subseteq B \subseteq V$, then $B^\perp \preceq A^\perp$.*

Definition 5.2.12. When a nonzero vector x in an inner product space V is divided by its norm the resulting vector $u = \dfrac{x}{\|x\|}$ is clearly a unit vector. We say that u results from NORMALIZING the vector x. A subset E of V is ORTHONORMAL if every pair of distinct vectors in E are orthogonal and every vector in E has length one. If, in addition, V is the span of E, then E is an ORTHONORMAL BASIS for V.

Definition 5.2.13. Let V be an inner product space and $E = \{e^1, e^2, \ldots, e^n\}$ be a finite orthonormal subset of V. For each $k \in \mathbb{N}_k$ let $x_k := \langle x, e^k \rangle$. This scalar is called the FOURIER COEFFICIENT of x with respect to E. The vector $s := \sum_{k=1}^n x_k e^k$ is the FOURIER SUM of x with respect to E.

Proposition 5.2.14. *Let notation be as in 5.2.13. Then $s = x$ if and only if $x \in \operatorname{span} E$.*

Proposition 5.2.15. *Let notation be as in 5.2.13. Then $x - s \perp e^k$ for $k = i, \ldots, n$ and therefore $x - s \perp s$.*

The next result gives us a recipe for converting a finite linearly independent subset of an inner product space into an orthonormal basis for the span of that set.

Theorem 5.2.16 (Gram-Schmidt Orthonormalization). *Let $A = \{a^1, a^2, \ldots, a^n\}$ be a finite linearly independent subset of an inner product space V. Define vectors $e^1, \ldots, e^n$ recursively by setting*

$$e^1 := \|a^1\|^{-1} a^1$$

and for $2 \leq m \leq n$

$$e^m := \|a^m - s^m\|^{-1}(a^m - s^m)$$

where $s^m := \sum_{k=1}^{m-1} \langle a^m, e^k \rangle e^k$ is the Fourier sum for a^m with respect to $E_{m-1} := \{e^1, \ldots, e^{m-1}\}$. Then E_n is an orthonormal basis for the span of A.

It should be clear from the proof of the preceding theorem that finiteness plays no essential role. The theorem remains true for countable linearly independent sets (and so does its proof).

Corollary 5.2.17. *Every finite dimensional inner product space has an orthonormal basis.*

Example 5.2.18. Let $\mathbb{R}[x]$ be the inner product space of real polynomials whose inner product is defined by

$$\langle p, q \rangle := \int_{-1}^{1} \widetilde{p}(x)\widetilde{q}(x)\, dx$$

for all $p,\ q \in \mathbb{R}[x]$. Application of the *Gram-Schmidt* process to the set $\{1, x, x^2, x^3, \ldots\}$ of real polynomials produces an orthonormal sequence of polynomials known as the *Legendre polynomials*. Compute the first four of these.

Example 5.2.19. Let $\mathbb{R}[x]$ be the inner product space of real polynomials whose inner product is defined by

$$\langle p, q \rangle := \int_0^\infty \widetilde{p}(x)\widetilde{q}(x)e^{-x}\,dx$$

for all p, $q \in \mathbb{R}[x]$. Application of the *Gram-Schmidt* process to the set $\{1, x, x^2, x^3, \ldots\}$ of real polynomials produces an orthonormal sequence of polynomials known as the *Laguerre polynomials*. Compute the first four of these.

Hint for proof. Integration by parts or familiarity with the gamma function allows us to conclude that $\int_0^\infty x^n e^{-x}\,dx = n!$ for each $n \in \mathbb{N}$.

Proposition 5.2.20. *If M is a subspace of a finite dimensional inner product space V then $V = M \oplus M^\perp$.*

Example 5.2.21. The subspace $l_c(\mathbb{N}, \mathbb{R})$ of the inner product space $l_2(\mathbb{R})$ (see Example 5.1.10) shows that the preceding proposition does not hold for infinite dimensional spaces.

Proposition 5.2.22. *Let M be a subspace of an inner product space V. Then*

(a) $M \subseteq M^{\perp\perp}$;
(b) *equality need not hold in* (a); *but*
(c) *if V is finite dimensional, then $M = M^{\perp\perp}$.*

Proposition 5.2.23. *If S is a set of mutually perpendicular vectors in an inner product space and $\mathbf{0} \notin S$, then the set S is linearly independent.*

Proposition 5.2.24. *Let M and N be subspaces of an inner product space V. Then*

(a) $(M + N)^\perp = (M \cup N)^\perp = M^\perp \cap N^\perp$ *and*
(b) *if V is finite dimensional, then* $(M \cap N)^\perp = M^\perp + N^\perp$.

Proposition 5.2.25. *Let S, $T\colon H \to K$ be linear maps between inner product spaces H and K. If $\langle Sx, y \rangle = \langle Tx, y \rangle$ for every $x \in H$ and $y \in K$, then $S = T$.*

Proposition 5.2.26. *If H is a complex inner product space and $T \in \mathfrak{L}(V)$ satisfies $\langle T\mathbf{z}, \mathbf{z} \rangle = 0$ for all $\mathbf{z} \in H$, then $T = 0$.*

Hint for proof. In the hypothesis replace $\mathbf{z}$ first by $\mathbf{x+y}$ and then by $\mathbf{x}+i\mathbf{y}$.

Example 5.2.27. Give an example to show that the preceding result does not hold for real inner product spaces.

Example 5.2.28. Let H be a complex inner product space and $a \in H$. Define $\psi_a : H \to \mathbb{C}$ by $\psi_a(x) = \langle x, a \rangle$ for all $x \in H$. Then ψ_a is a linear functional on H.

Theorem 5.2.29 (Riesz-Fréchet Theorem). *If f is a linear functional on a finite dimensional inner product space H, then there exists a unique vector $a \in H$ such that*

$$f(x) = \langle x, a \rangle$$

for every $x \in H$.

Example 5.2.30. Consider the function $\phi : l_c(\mathbb{N}) \to \mathbb{C} : x \mapsto \sum_{k=1}^{\infty} \alpha_k$ where $x = \sum_{k=1}^{\infty} \alpha_k e^k$, the e^k's being the standard basis vectors for $l_c(\mathbb{N})$. This linear functional provides an example which shows that the *Riesz-Fréchet theorem* does not hold (as stated) in infinite dimensional spaces.

Here is another example of the failure of this result in infinite dimensional spaces.

Example 5.2.31. On the vector space H of polynomials over $\mathbb{C}$ define an inner product by $\langle p, q \rangle = \int_0^1 p(t)\overline{q(t)}\, dt$. For a fixed $z \in \mathbb{C}$ define the functional E_z, called *evaluation at z*, by $E_z(p) = p(z)$ for every $p \in H$. Then E_z belongs to H^* but that there is no polynomial p such that $E_z(q) = \langle q, p \rangle$ for every $q \in H$.

Caution 5.2.32. It is important not to misinterpret the two preceding examples. There is indeed a (very important!) version of the *Riesz-Fréchet theorem* that does in fact hold for infinite dimensional spaces. If we restrict our attention to *continuous* linear functionals on *complete* inner product spaces, then the conclusion of Theorem 5.2.29 does indeed hold for infinite dimensional spaces.

Exercise 5.2.33. Use vector methods (as described in 1.4.15 — no coordinates, no major results from Euclidean geometry) to show that the midpoint of the hypotenuse of a right triangle is equidistant from the vertices. *Hint.* Let $\triangle ABC$ be a right triangle and O be the midpoint of the hypotenuse AB. What can you say about $\langle \overrightarrow{AO} + \overrightarrow{OC}, \overrightarrow{CO} + \overrightarrow{OB} \rangle$?

Exercise 5.2.34. Use vector methods (as described in 1.4.15) to show that an angle inscribed in a semicircle is a right angle.

Exercise 5.2.35. Use vector methods (as described in 1.4.15) to show that if a parallelogram has perpendicular diagonals, then it is a rhombus (that is, all four sides have equal length). *Hint.* Let $ABCD$ be a parallelogram. Express the inner product of the diagonals $\overrightarrow{AC}$ and $\overrightarrow{DB}$ in terms of the lengths of the sides $\overrightarrow{AB}$ and $\overrightarrow{BC}$.

Exercise 5.2.36. Use vector methods (as described in 1.4.15) to show that the diagonals of a rhombus are perpendicular.

5.3. Involutions and Adjoints

Definition 5.3.1. An INVOLUTION on a complex (or real) algebra A is a map $x \mapsto x^*$ from A into A that satisfies

(i) $(x + y)^* = x^* + y^*$,
(ii) $(\alpha x)^* = \overline{\alpha}\, x^*$,
(iii) $x^{**} = x$, and
(iv) $(xy)^* = y^* x^*$

for all $x, y \in A$ and $\alpha \in \mathbb{C}$ (or $\mathbb{R}$). An algebra on which an involution has been defined is a $*$-ALGEBRA (pronounced "star algebra"). An algebra homomorphism ϕ between $*$-algebras which preserves involution (that is, such that $\phi(a^*) = (\phi(a))^*$) is a $*$-HOMOMORPHISM (pronounced "star homomorphism"). A $*$-homomorphism $\phi \colon A \to B$ between unital algebras is said to be UNITAL if $\phi(1_A) = 1_B$.

Proposition 5.3.2. *If a and b are elements of a $*$-algebra, then a commutes with b if and only if a^* commutes with b^*.*

Proposition 5.3.3. *In a unital $*$-algebra $1^* = 1$.*

Proposition 5.3.4. *If a $*$-algebra A has a left multiplicative identity e, then A is unital and $e = 1_A$.*

Proposition 5.3.5. *An element a of a unital $*$-algebra is invertible if and only if a^* is. And when a is invertible we have*

$$(a^*)^{-1} = (a^{-1})^*.$$

Proposition 5.3.6. *Let a be an element of a unital $*$-algebra. Then $\lambda \in \sigma(a)$ if and only if $\overline{\lambda} \in \sigma(a^*)$.*

Definition 5.3.7. An element a of a complex $*$-algebra A is NORMAL if $a^*a = aa^*$. It is SELF-ADJOINT (or HERMITIAN) if $a^* = a$. It is SKEW-HERMITIAN if $a^* = -a$. And it is UNITARY if $a^*a = aa^* = 1$. The set of all self-adjoint elements of A is denoted by $\mathcal{H}(A)$, the set of all normal elements by $\mathcal{N}(A)$, and the set of all unitary elements by $\mathcal{U}(A)$.

Oddly, and perhaps somewhat confusingly, history has dictated an alternative, but parallel, language for real algebras — especially algebras of matrices and linear maps. An element a of a real $*$-algebra A is SYMMETRIC if $a^* = a$. It is SKEW-SYMMETRIC if $a^* = -a$. And it is ORTHOGONAL if $a^*a = aa^* = 1$.

Example 5.3.8. Complex conjugation is an involution on the algebra $\mathbb{C}$ of complex numbers.

Example 5.3.9. Transposition (see Definition 1.7.31) is an involution on the real algebra $\mathbf{M}_n$ of $n \times n$ matrices.

Example 5.3.10. Let $a < b$. The map $f \mapsto \overline{f}$ of a function to its complex conjugate is an involution on the complex algebra $\mathcal{C}([a,b], \mathbb{C})$ of continuous complex valued functions on $[a,b]$.

Proposition 5.3.11. *For every element a of a $*$-algebra A there exist unique self-adjoint elements u and v in A such that $a = u + iv$.*

Hint for proof. The self-adjoint element u is called the *real part* of a and v the *imaginary part* of a.

Corollary 5.3.12. *An element of a $*$-algebra is normal if and only if its real part and its imaginary part commute.*

Definition 5.3.13. Let H and K be complex inner product spaces and $T \colon H \to K$ be a linear map. If there exists a function $T^* \colon K \to H$ that satisfies

$$\langle Tx, y \rangle = \langle x, T^*y \rangle$$

for all $x \in H$ and $y \in K$, then T^* is the ADJOINT of T. If a linear map T has an adjoint we say that T is ADJOINTABLE. Denote the set of all adjointable maps from H to K by $\mathfrak{A}(H, K)$ and write $\mathfrak{A}(H)$ for $\mathfrak{A}(H, H)$.

When H and K are real vector spaces, the adjoint of T is usually called the TRANSPOSE of T and the notation T^t is used (rather than T^*).

Proposition 5.3.14. *Let $T: H \to K$ be a linear map between complex inner product spaces. If the adjoint of T exists, then it is unique. (That is, there is at most one function $T^*: K \to H$ that satisfies $\langle Tx, y \rangle = \langle x, T^*y \rangle$ for all $x \in H$ and $y \in K$.)*

Similarly, of course, if $T: H \to K$ is a linear map between real inner product spaces and if the transpose of T exists, then it is unique.

Example 5.3.15. Let $\mathcal{C} = \mathcal{C}([0,1])$ be the inner product space defined in Example 5.1.4 and $J_0 = \{f \in \mathcal{C}: f(0) = 0\}$. Then the inclusion map $\iota: J_0 \to \mathcal{C}$ is an example of a map that is *not* adjointable.

Example 5.3.16. Let U be the unilateral shift operator on l_2 (see Example 5.1.10)

$$U: l_2 \to l_2: (x_1, x_2, x_3, \ldots) \mapsto (0, x_1, x_2, \ldots),$$

then its adjoint is given by

$$U^*: l_2 \to l_2: (x_1, x_2, x_3, \ldots) \mapsto (x_2, x_3, x_4, \ldots).$$

Example 5.3.17 (Multiplication operators). Let ϕ be a fixed continuous complex valued function on the interval $[a, b]$. On the inner product space $\mathcal{C} = \mathcal{C}([a, b], \mathbb{C})$ (see Example 5.1.4) define

$$M_\phi: \mathcal{C} \to \mathcal{C}: f \mapsto \phi f.$$

Then M_ϕ is an adjointable operator on $\mathcal{C}$.

Proposition 5.3.18. *Let $T: H \to K$ be a linear map between complex inner product spaces. If the adjoint of T exists, then it is linear.*

And, similarly, of course, if the transpose of a linear map between real inner product spaces exists, then it is linear. In the sequel we will forgo the dubious helpfulness of mentioning every obvious real analog of results holding for complex inner product spaces.

Proposition 5.3.19. *Let $T: H \to K$ be a linear map between complex inner product spaces. If the adjoint of T exists, then so does the adjoint of T^* and $T^{**} = T$.*

Proposition 5.3.20. *Let $S: H \to K$ and $T: K \to L$ be linear maps between complex inner product space. Show that if S and T both have adjoints, then so does their composite TS and*

$$(TS)^* = S^*T^*.$$

Proposition 5.3.21. *If $T: H \to K$ is an invertible linear map between complex inner product spaces and both T and T^{-1} have adjoints, then T^* is invertible and $(T^*)^{-1} = (T^{-1})^*$.*

Proposition 5.3.22. *Let S and T be operators on a complex inner product space H. Then $(S + T)^* = S^* + T^*$ and $(\alpha T)^* = \overline{\alpha} T^*$ for every $\alpha \in \mathbb{C}$.*

Example 5.3.23. If H is a complex inner product space, then $\mathfrak{A}(H)$ is a unital complex $*$-algebra. It is a unital subalgebra of $\mathfrak{L}(H)$.

Theorem 5.3.24. *Let T be an adjointable operator on an inner product space H. Then*

(a) $\ker T = (\operatorname{ran} T^*)^{\perp}$ *and*
(b) $\operatorname{ran} T^* \subseteq (\ker T)^{\perp}$. *If H is finite dimensional, then equality holds in* (b).

Theorem 5.3.25. *Let T be an adjointable operator on an inner product space H. Then*

(a) $\ker T^* = (\operatorname{ran} T)^{\perp}$ *and*
(b) $\operatorname{ran} T \subseteq (\ker T^*)^{\perp}$. *If H is finite dimensional, then equality holds in* (b).

Proposition 5.3.26. *Every linear map between finite dimensional complex inner product spaces is adjointable.*

Hint for proof. Use the *Riesz-Fréchet theorem.*

Corollary 5.3.27. *If H is a finite dimensional inner product space then $\mathfrak{L}(H)$ is a unital $*$-algebra.*

Proposition 5.3.28. *An adjointable operator T on a complex inner product space H is unitary if and only if it preserves inner products (that is, if and only if $\langle Tx, Ty \rangle = \langle x, y \rangle$ for all x, $y \in H$). Similarly, an adjointable operator on a real inner product space is orthogonal if and only if it preserves inner products.*

Exercise 5.3.29. Let $T: H \to K$ be a linear map between finite dimensional complex inner product spaces. Find the matrix representation of T^* in terms of the matrix representation of T. Also, for a linear map $T: H \to K$ between finite dimensional real inner product spaces find the matrix representation of T^t in terms of the matrix representation of T.

Proposition 5.3.30. *Every eigenvalue of a Hermitian operator on a complex inner product space is real.*

Hint for proof. Let x be an eigenvector associated with an eigenvalue λ of an operator A. Consider $\lambda \|x\|^2$.

Proposition 5.3.31. *Let A be a Hermitian operator on a complex inner product space. Prove that eigenvectors associated with distinct eigenvalues of A are orthogonal.*

Hint for proof. Let x and y be eigenvectors associated with distinct eigenvalues λ and μ of A. Start your proof by showing that $\lambda\langle x, y\rangle = \mu\langle x, y\rangle$.

Proposition 5.3.32. *Let N be a normal operator on a complex inner product space H. Then $\|Nx\| = \|N^*x\|$ for every $x \in H$.*

5.4. Orthogonal Projections

Proposition 5.4.1. *Let H be an inner product space and M and N be subspaces of H such that $H = M + N$ and $M \cap N = \{0\}$. (That is, suppose that H is the* vector space *direct sum of M and N.) Also let $P = E_{NM}$ be the projection of H along N onto M. Prove that P is self-adjoint (P^* exists and $P^* = P$) if and only if $M \perp N$.*

Definition 5.4.2. A PROJECTION in a $*$-algebra A is an element p of the algebra that is idempotent ($p^2 = p$) and self-adjoint ($p^* = p$). The set of all projections in A is denoted by $\mathcal{P}(A)$.

An operator P on a complex inner product space H is an ORTHOGONAL PROJECTION if it is self-adjoint and idempotent; that is, if it is a projection in the $*$-algebra $\mathfrak{A}(H)$ of adjointable operators on H. (On a real inner product space, of course, the appropriate adjectives are *symmetric* and *idempotent*.)

Notice that a vector space projection E_{MN} is, in general, *not* a projection in any $*$-algebra.

Convention 5.4.3. It is standard practice to refer to orthogonal projections on inner product spaces simply as "projections". This clearly invites confusion with the vector space notion of projection. So one must be careful: just as the symbols $\oplus$ and $\perp$ have different meanings depending on context (vector spaces or inner product spaces), so does the word "projection". In these notes and elsewhere, when the context is inner product spaces the word "projection" frequently means "orthogonal projection".

Proposition 5.4.4. *If P is an orthogonal projection on an inner product space, then the space is the orthogonal direct sum of the range of P and the kernel of P.*

Proposition 5.4.5. *Let p and q be projections in a $*$-algebra. Then the following are equivalent:*

(a) $pq = \mathbf{0}$;
(b) $qp = \mathbf{0}$;
(c) $qp = -pq$;
(d) $p + q$ *is a projection.*

Definition 5.4.6. Let p and q be projections in a $*$-algebra. If any of the conditions in the preceding result holds, then we say that p and q are ORTHOGONAL and write $p \perp q$. (Thus for operators on an inner product space we can correctly speak of orthogonal projections!)

Proposition 5.4.7. *Let P and Q be projections on an inner product space H. Then $P \perp Q$ if and only if $\operatorname{ran} P \perp \operatorname{ran} Q$. In this case $P + Q$ is an orthogonal projection whose kernel is $\ker P \cap \ker Q$ and whose range is $\operatorname{ran} P + \operatorname{ran} Q$.*

Example 5.4.8. On an inner product space projections need not commute. For example, let P be the projection of the (real) inner product space R^2 onto the line $y = x$ and Q be the projection of R^2 onto the x-axis. Then $PQ \neq QP$.

Proposition 5.4.9. *Let p and q be projections in a $*$-algebra. Then pq is a projection if and only if $pq = qp$.*

Proposition 5.4.10. *Let P and Q be projections on an inner product space H. If $PQ = QP$, then PQ is a projection whose kernel is $\ker P + \ker Q$ and whose range is $\operatorname{ran} P \cap \operatorname{ran} Q$.*

Proposition 5.4.11. *Let p and q be projections in a $*$-algebra. Then the following are equivalent:*

(a) $pq = p$;
(b) $qp = p$;
(c) $q - p$ *is a projection.*

Definition 5.4.12. Let p and q be projections in a $*$-algebra. If any of the conditions in the preceding result holds, then we write $p \preceq q$. In this case we say that p is a SUBPROJECTION of q or that p is SMALLER than q.

Proposition 5.4.13. *If A is a $*$-algebra, then the relation $\preceq$ defined in 5.4.12 is a partial ordering on $\mathcal{P}(A)$. If A is unital, then $\mathbf{0} \preceq p \preceq \mathbf{1}$ for every $p \in \mathcal{P}(A)$.*

Notation 5.4.14. If H, M, and N are subspaces of an inner product space, then the assertion $H = M \oplus N$, may be rewritten as $M = H \ominus N$ (or $N = H \ominus M$).

Proposition 5.4.15. *Let P and Q be projections on an inner product space H. Then the following are equivalent:*

(a) $P \preceq Q$;
(b) $\|Px\| \leq \|Qx\|$ *for all $x \in H$; and*
(c) $\operatorname{ran} P \subseteq \operatorname{ran} Q$.

In this case $Q - P$ is a projection whose kernel is $\operatorname{ran} P + \ker Q$ and whose range is $\operatorname{ran} Q \ominus \operatorname{ran} P$.

The next two results are optional: they will not be used in the sequel.

Proposition 5.4.16. *Suppose p and q are projections on a $*$-algebra A. If $pq = qp$, then the infimum of p and q, which we denote by $p \curlywedge q$, exists with respect to the partial ordering $\preceq$ and $p \curlywedge q = pq$. The infimum $p \curlywedge q$ may exist even when p and q do not commute. A necessary and sufficient condition that $p \perp q$ hold is that both $p \curlywedge q = \mathbf{0}$ and $pq = qp$ hold.*

Proposition 5.4.17. *Suppose p and q are projections on a $*$-algebra A. If $p \perp q$, then the supremum of p and q, which we denote by $p \curlyvee q$, exists with respect to the partial ordering $\preceq$ and $p \curlyvee q = p + q$. The supremum $p \curlyvee q$ may exist even when p and q are not orthogonal.*

5.5. The Spectral Theorem for Inner Product Spaces

Definition 5.5.1. Two elements a and b of a *-algebra A are UNITARILY EQUIVALENT if there exists a unitary element u of A such that $b = u^*au$.

Definition 5.5.2. An operator T on a complex inner product space V is UNITARILY DIAGONALIZABLE if there exists an orthonormal basis for V consisting of eigenvectors of T.

Proposition 5.5.3. *Let A be an $n \times n$ matrix A of complex numbers. Then A, regarded as an operator on $\mathbb{C}^n$, is unitarily diagonalizable if and only if it is unitarily equivalent to a diagonal matrix.*

Definition 5.5.4. Let $M_1 \oplus \cdots \oplus M_n$ be an orthogonal direct sum decomposition of an inner product space H. For each k let P_k be the orthogonal projection onto M_k. The projections $P_1, \ldots, P_n$ are the ORTHOGONAL PROJECTIONS ASSOCIATED WITH THE ORTHOGONAL DIRECT SUM DECOMPOSITION $H = M_1 \oplus \cdots \oplus M_n$. The family $\{P_1, \ldots, P_n\}$ is an ORTHOGONAL RESOLUTION OF THE IDENTITY. (Compare this with Definitions 4.9.1 and 4.9.2.)

Theorem 5.5.5 (Spectral Theorem: Complex Inner Product Space Version). *An operator N on a finite dimensional complex inner product space V is normal if and only if it is unitarily diagonalizable in which case it can be written as*

$$N = \sum_{k=1}^{n} \lambda_k P_k$$

where $\lambda_1, \ldots, \lambda_n$ are the (distinct) eigenvalues of N and $\{P_1, \ldots, P_n\}$ is the orthogonal resolution of the identity whose orthogonal projections are associated with the corresponding eigenspaces $M_1, \ldots, M_n$.

Proof. See [29], Theorems 10.13 and 10.21; or [16], Chapter 8, Theorems 20 and 22, and Chapter 9, Theorem 9.

Exercise 5.5.6. Let N be the operator on $\mathbb{C}^2$ whose matrix representation is

$$\begin{bmatrix} 0 & 1 \\ -1 & 0 \end{bmatrix}.$$

(a) The eigenspace M_1 associated with the eigenvalue $-i$ is the span of $(1, \underline{\hspace{1cm}})$.

(b) The eigenspace M_2 associated with the eigenvalue i is the span of $(1, \underline{\hspace{1cm}})$.

(c) The (matrix representations of the) orthogonal projections P_1 and P_2 onto the eigenspaces M_1 and M_2, respectively, are $P_1 = \begin{bmatrix} a & b \\ -b & a \end{bmatrix}$; and $P_2 = \begin{bmatrix} a & -b \\ b & a \end{bmatrix}$ where $a = \underline{\hspace{1cm}}$ and $b = \underline{\hspace{1cm}}$.

(d) Write N as a linear combination of the projections found in (c).

Answer: $[N] = $ _____ $P_1 + $ _____ P_2.

(e) A unitary matrix U that diagonalizes $[N]$ is $\begin{bmatrix} a & a \\ -b & b \end{bmatrix}$ where $a = $ _____ and $b = $ _____ .

The associated diagonal form $\Lambda = U^*[N]U$ of $[N]$ is $\begin{bmatrix} & \\ & \end{bmatrix}$.

Exercise 5.5.7. Let H be the self-adjoint matrix $\begin{bmatrix} 2 & 1+i \\ 1-i & 3 \end{bmatrix}$.

(a) Use the *spectral theorem* to write H as a linear combination of orthogonal projections.

Answer: $H = \alpha P_1 + \beta P_2$ where $\alpha = $ _____ , $\beta = $ _____ , $P_1 = \frac{1}{3}\begin{bmatrix} 2 & -1-i \\ & \end{bmatrix}$, and $P_2 = \frac{1}{3}\begin{bmatrix} 1 & 1+i \\ & \end{bmatrix}$.

(b) Find a square root of H.

Answer: $\sqrt{H} = \frac{1}{3}\begin{bmatrix} 4 & 1+i \\ & \end{bmatrix}$.

Exercise 5.5.8. Let $N = \frac{1}{3}\begin{bmatrix} 4+2i & 1-i & 1-i \\ 1-i & 4+2i & 1-i \\ 1-i & 1-i & 4+2i \end{bmatrix}$.

(a) The matrix N is normal because $NN^* = N^*N = \begin{bmatrix} a & b & b \\ b & a & b \\ b & b & a \end{bmatrix}$ where $a = $ _____ and $b = $ _____ .

(b) According to the *spectral theorem* N can be written as a linear combination of orthogonal projections. Written in this form $N = \lambda_1 P_1 + \lambda_2 P_2$ where $\lambda_1 = $ _____ ,

$\lambda_2 = $ _____ , $P_1 = \begin{bmatrix} a & a & a \\ a & a & a \\ a & a & a \end{bmatrix}$, and $P_2 = \begin{bmatrix} b & -a & -a \\ -a & b & -a \\ -a & -a & b \end{bmatrix}$ where

$a = $ _____ and $b = $ _____ .

(c) A unitary matrix U that diagonalizes N is $\begin{bmatrix} a & -b & -c \\ a & b & -c \\ a & d & 2c \end{bmatrix}$ where

$a = \underline{\hspace{1cm}}$, $b = \underline{\hspace{1cm}}$, $c = \underline{\hspace{1cm}}$, and $d = \underline{\hspace{1cm}}$.

The associated diagonal form $\Lambda = U^* N U$ of N is $\begin{bmatrix} & & \\ & & \\ & & \end{bmatrix}$.

Chapter 6

A BRIEF REVIEW OF DIFFERENTIAL CALCULUS

We now pause for a *very* brief review of differential calculus. The central concept here is differentiability. A function f between normed linear spaces is said to be *differentiable* at a point p if (when the point $(p, f(p))$ is translated to the origin) the function is tangent to some continuous linear map. In this chapter (much of which is just Chapter 13 of my online text [11]) we make this idea precise and record a few important facts about differentiability. A more detailed and leisure treatment can be found in my *A Problems Based Course in Advanced Calculus* [12], Chapters 25–30.

There are two sorts of textbooks on differential calculus: concept oriented and computation oriented. It is my belief that students who understand the concepts behind differentiation can do the calculations, while students who study calculations only often get stuck. Among the most masterful presentations of concept oriented differential calculus are [8] (Volume I, Chapter 8) and [22] (Chapter 3). As of this writing the latter book is available without charge at the website of one of the authors:

http://www.math.harvard.edu/~shlomo/docs/Advanced_Calculus.pdf

The material in this chapter will benefit primarily those whose only encounter with multivariate calculus has been through partial derivatives and a *chain rule* that looks something like

$$\frac{\partial w}{\partial u} = \frac{\partial w}{\partial x}\frac{\partial x}{\partial u} + \frac{\partial w}{\partial y}\frac{\partial y}{\partial u} + \frac{\partial w}{\partial z}\frac{\partial z}{\partial u} \tag{6.1}$$

The approach here is intended to be more geometric, emphasizing the role of *tangency*.

6.1. Tangency

Notation 6.1.1. Let V and W be normed linear spaces and $a \in V$. (If you are unfamiliar with, or uncomfortable working in, normed linear spaces, just pretend that all the spaces involved are n-dimensional Euclidean spaces. The only thing you may lose by doing so is the pleasant feeling of assurance that differential calculus is no harder in infinite dimensional spaces than on the real line.) We denote by $\mathcal{F}_a(V, W)$ the family of all functions defined on a neighborhood of a taking values in W. That is, f belongs to $\mathcal{F}_a(V, W)$ if there exists an open set U such that $a \in U \subseteq \operatorname{dom} f \subseteq V$ and if the image of f is contained in W. We shorten $\mathcal{F}_a(V, W)$ to $\mathcal{F}_a$ when no confusion will result. Notice that for each $a \in V$, the set $\mathcal{F}_a$ is closed under addition and scalar multiplication. (As usual, we define the sum of two functions f and g in $\mathcal{F}_a$ to be the function $f + g$ whose value at x is $f(x) + g(x)$ whenever x belongs to $\operatorname{dom} f \cap \operatorname{dom} g$.) Despite the closure of $\mathcal{F}_a$ under these operations, $\mathcal{F}_a$ is *not* a vector space. (Why not?)

Definition 6.1.2. Let V and W be normed linear spaces. A function f in $\mathcal{F}_0(V, W)$ belongs to $\mathfrak{O}(V, W)$ if there exist numbers $c > 0$ and $\delta > 0$ such that

$$\|f(x)\| \leq c \, \|x\|$$

whenever $\|x\| < \delta$.

A function f in $\mathcal{F}_0(V, W)$ belongs to $\mathfrak{o}(V, W)$ if for every $c > 0$ there exists $\delta > 0$ such that

$$\|f(x)\| \leq c \, \|x\|$$

whenever $\|x\| < \delta$. Notice that f belongs to $\mathfrak{o}(V, W)$ if and only if $f(0) = 0$ and

$$\lim_{h \to 0} \frac{\|f(h)\|}{\|h\|} = 0.$$

When no confusion seems likely we will shorten $\mathfrak{O}(V, W)$ to $\mathfrak{O}$ and $\mathfrak{o}(V, W)$ to $\mathfrak{o}$.

Exercise 6.1.3. Here is a list summarizing the important facts about the families $\mathfrak{D}$ and $\mathfrak{o}$. State precisely what each of these says and give a proof. (Here $\mathfrak{B}$ is the set of continuous linear maps between normed linear spaces V and W, and $\mathcal{C}_0$ is the set of all functions in $\mathcal{F}_0(V, W)$ that are continuous at 0.)

$$(1) \qquad \mathfrak{B} \cup \mathfrak{o} \subseteq \mathfrak{D} \subseteq \mathcal{C}_0.$$

$$(2) \qquad \mathfrak{B} \cap \mathfrak{o} = 0.$$

$$(3) \qquad \mathfrak{D} + \mathfrak{D} \subseteq \mathfrak{D}; \qquad \alpha \mathfrak{D} \subseteq \mathfrak{D}.$$

$$(4) \qquad \mathfrak{o} + \mathfrak{o} \subseteq \mathfrak{o}; \qquad \alpha \mathfrak{o} \subseteq \mathfrak{o}.$$

$$(5) \qquad \mathfrak{o} \circ \mathfrak{D} \subseteq \mathfrak{o}.$$

$$(6) \qquad \mathfrak{D} \circ \mathfrak{o} \subseteq \mathfrak{o}.$$

$$(7) \qquad \mathfrak{o}(V, \mathbb{R}) \cdot W \subseteq \mathfrak{o}(V, W).$$

$$(8) \qquad \mathfrak{D}(V, \mathbb{R}) \cdot \mathfrak{D}(V, W) \subseteq \mathfrak{o}(V, W).$$

Definition 6.1.4. Let V and W be normed linear spaces. Two functions f and g in $\mathcal{F}_0(V, W)$ are TANGENT (AT ZERO), in which case we write $f \simeq g$, if $f - g \in \mathfrak{o}(V, W)$.

Proposition 6.1.5. *The relation of tangency enjoys the following properties.*

(a) *"Tangency at zero" is an equivalence relation on $\mathcal{F}_0$.*
(b) *Let $S, T \in \mathfrak{B}$ and $f \in \mathcal{F}_0$. If $S \simeq f$ and $T \simeq f$, then $S = T$.*
(c) *If $f \simeq g$ and $j \simeq k$, then $f + j \simeq g + k$, and furthermore, $\alpha f \simeq \alpha g$ for all $\alpha \in \mathbb{R}$.*
(d) *Let $\phi, \psi \in \mathcal{F}_0(V, \mathbb{R})$ and $w \in W$. If $\phi \simeq \psi$, then $\phi w \simeq \psi w$.*
(e) *Let $f, g \in \mathcal{F}_0(V, W)$ and $T \in \mathfrak{B}(W, X)$. If $f \simeq g$, then $T \circ f \simeq T \circ g$.*
(f) *Let $h \in \mathfrak{D}(V, W)$ and $f, g \in \mathcal{F}_0(W, X)$. If $f \simeq g$, then $f \circ h \simeq g \circ h$.*

6.2. The Differential

Definition 6.2.1. Let V and W be normed linear spaces, $a \in V$, and $f \in \mathcal{F}_a(V, W)$. Define the function Δf_a by

$$\Delta f_a(h) := f(a + h) - f(a)$$

for all h such that $a + h$ is in the domain of f. Notice that since f is defined in a neighborhood of a, the function Δf_a is defined in a neighborhood of 0; that is, Δf_a belongs to $\mathcal{F}_0(V, W)$. Notice also that $\Delta f_a(0) = 0$.

Proposition 6.2.2. *If V and W are normed linear spaces and $a \in V$, then the function Δ has the following properties.*

(a) *If $f \in \mathcal{F}_a(V, W)$ and $\alpha \in \mathbb{R}$, then*

$$\Delta(\alpha f)_a = \alpha \, \Delta f_a.$$

(b) *If $f, g \in \mathcal{F}_a(V, W)$, then*

$$\Delta(f + g)_a = \Delta f_a + \Delta g_a.$$

(c) *If $\phi \in \mathcal{F}_a(V, \mathbb{R})$ and $f \in \mathcal{F}_a(V, W)$, then*

$$\Delta(\phi f)_a = \phi(a) \cdot \Delta f_a + \Delta \phi_a \cdot f(a) + \Delta \phi_a \cdot \Delta f_a.$$

(d) *If $f \in \mathcal{F}_a(V, W)$, $g \in \mathcal{F}_{f(a)}(W, X)$, and $g \circ f \in \mathcal{F}_a(V, X)$, then*

$$\Delta(g \circ f)_a = \Delta g_{f(a)} \circ \Delta f_a.$$

(e) *A function $f \colon V \to W$ is continuous at the point a in V if and only if Δf_a is continuous at 0.*

(f) *If $f \colon U \to U_1$ is a bijection between subsets of arbitrary vector spaces, then for each a in U the function $\Delta f_a \colon U - a \to U_1 - f(a)$ is invertible and*

$$\left(\Delta f_a\right)^{-1} = \Delta\left(f^{-1}\right)_{f(a)}.$$

Definition 6.2.3. Let V and W be normed linear spaces, $a \in V$, and $f \in \mathcal{F}_a(V, W)$. We say that f is DIFFERENTIABLE AT a if there exists a continuous linear map that is tangent at 0 to Δf_a. If such a map exists, it is called the DIFFERENTIAL of f at a and is denoted by df_a. Thus df_a is just a member of $\mathfrak{B}(V, W)$ that satisfies $df_a \simeq \Delta f_a$. We denote by $\mathcal{D}_a(V, W)$ the family of all functions in $\mathcal{F}_a(V, W)$ that are differentiable at a. We often shorten this to $\mathcal{D}_a$.

We establish next that there can be at most one bounded linear map tangent to Δf_a.

Proposition 6.2.4. *Let V and W be normed linear spaces and $a \in V$. If $f \in \mathcal{D}_a(V, W)$, then its differential is unique.*

Exercise 6.2.5. Let

$$f \colon \mathbb{R}^3 \to \mathbb{R}^2 \colon (x, y, z) \mapsto (x^2 y - 7, 3xz + 4y)$$

and $a = (1, -1, 0)$. Use the *definition* of "differential" to find df_a. *Hint.* Work with the matrix representation of df_a. Since the differential must belong to $\mathfrak{B}(\mathbb{R}^3, \mathbb{R}^2)$, its matrix representation is a 2×3 matrix $M = \begin{bmatrix} r & s & t \\ u & v & w \end{bmatrix}$. Use the requirement that $\|h\|^{-1} \|\Delta f_a(h) - Mh\| \to 0$ as $h \to 0$ to discover the identity of the entries in M.

Exercise 6.2.6. Let $\mathbf{F} \colon \mathbb{R}^2 \to \mathbb{R}^4$ be defined by $\mathbf{F}(x, y) = (y, x^2, 4 - xy, 7x)$, and let $\mathbf{p} = (1, 1)$. Use the *definition* of "differentiable" to show that $\mathbf{F}$ is differentiable at $\mathbf{p}$. Find the (matrix representation of the) differential of $\mathbf{F}$ at p.

Proposition 6.2.7. *Let V and W be normed linear spaces and $a \in V$. If $f \in \mathcal{D}_a$, then $\Delta f_a \in \mathfrak{O}$; thus, every function that is differentiable at a point is continuous there.*

Proposition 6.2.8. *Let V and W be normed linear spaces and $a \in V$. Suppose that $f, g \in \mathcal{D}_a(V, W)$ and that $\alpha \in \mathbb{R}$. Then*

(1) αf *is differentiable at a and*

$$d(\alpha f)_a = \alpha \, df_a;$$

(2) *also, $f + g$ is differentiable at a and*

$$d(f + g)_a = df_a + dg_a.$$

Suppose further that $\phi \in \mathcal{D}_a(V, \mathbb{R})$. Then

(c) $\phi f \in \mathcal{D}_a(V, W)$ *and*

$$d(\phi f)_a = d\phi_a \cdot f(a) + \phi(a) \, df_a.$$

It seems to me that the version of the *chain rule* given in Equation (6.1) above, although (under appropriate hypotheses) a correct equation, really provides us with very little insight about what is going on. There is, well-hidden in the symbols, an important and straightforward concept that deserves to be clearly displayed; it is, quite simply, that the best linear approximation to the composite of two smooth functions is the composite of their best linear approximations.

Theorem 6.2.9 (The Chain Rule). *Let V, W, and X be normed linear spaces with $a \in V$. If $f \in \mathcal{D}_a(V, W)$ and $g \in \mathcal{D}_{f(a)}(W, X)$, then $g \circ f \in \mathcal{D}_a(V, X)$ and*

$$d(g \circ f)_a = dg_{f(a)} \circ df_a.$$

Proof. Our hypotheses are $\Delta f_a \simeq df_a$ and $\Delta g_{f(a)} \simeq dg_{f(a)}$. By Proposition 6.2.7 $\Delta f_a \in \mathfrak{O}$. Then by Proposition 6.1.5(f)

$$\Delta g_{f(a)} \circ \Delta f_a \simeq dg_{f(a)} \circ \Delta f_a \tag{6.2}$$

and by Proposition 6.1.5(e)

$$dg_{f(a)} \circ \Delta f_a \simeq dg_{f(a)} \circ df_a. \tag{6.3}$$

According to Proposition 6.2.2(d)

$$\Delta (g \circ f)_a \simeq \Delta g_{f(a)} \circ \Delta f_a. \tag{6.4}$$

From (6.2), (6.3), (6.4), and Proposition 6.1.5(a) it is clear that

$$\Delta (g \circ f)_a \simeq dg_{f(a)} \circ df_a.$$

Since $dg_{f(a)} \circ df_a$ is a bounded linear transformation, the desired conclusion is an immediate consequence of Proposition 6.2.4. $\qquad\qquad \square$

Exercise 6.2.10. Derive (under appropriate hypotheses) Equation (6.1) from Theorem 6.2.9.

Exercise 6.2.11. Let $\mathbf{T}$ be a linear map from $\mathbb{R}^n$ to $\mathbb{R}^m$ and $\mathbf{p} \in \mathbb{R}^n$. Find $d\mathbf{T_p}$.

Example 6.2.12. Let T be a symmetric $n \times n$ matrix and let $\mathbf{p} \in \mathbb{R}^n$. Define a function $f \colon \mathbb{R}^n \to \mathbb{R}$ by $f(\mathbf{x}) = \langle T\mathbf{x}, \mathbf{x} \rangle$. Then

$$df_{\mathbf{p}}(\mathbf{h}) = 2\langle T\mathbf{p}, \mathbf{h} \rangle$$

for every $h \in \mathbb{R}^n$.

6.3. The Gradient of a Scalar Field in $\mathbb{R}^n$

Definition 6.3.1. A SCALAR FIELD on $\mathbb{R}^n$ is a scalar valued function on a subset of $\mathbb{R}^n$.

Definition 6.3.2. Let U be an open subset of $\mathbb{R}^n$ and $\phi \colon U \to \mathbb{R}$ be a scalar field. If ϕ is differentiable at a point a in U, then its differential $d\phi_a$ is a (continuous) linear map from $\mathbb{R}^n$ into $\mathbb{R}$. That is, $d\phi_a \in (\mathbb{R}^n)^*$. Thus according to the *Riesz-Fréchet Theorem* 5.2.29 there exists a unique vector, which we denote by $\nabla\phi(a)$, representing the linear functional $d\phi_a$. That is, $\nabla\phi(a)$ is the unique vector in $\mathbb{R}^n$ such that

$$d\phi_a(x) = \langle x, \nabla\phi(a)\rangle$$

for all x in $\mathbb{R}^n$. The vector $\nabla\phi(a)$ is the GRADIENT of ϕ at a. If U is an open subset of $\mathbb{R}^n$ and ϕ is differentiable at each point of U, then the function

$$\nabla\phi \colon U \to \mathbb{R}^n \colon u \mapsto \nabla\phi(u)$$

is the GRADIENT of ϕ. Notice two things: first, the gradient of a scalar field is a vector field (that is, a map from $\mathbb{R}^n$ into $\mathbb{R}^n$); and second, the differential $d\phi_a$ is the zero linear functional if and only if the gradient at a, $\nabla\phi(a)$, is the zero vector in $\mathbb{R}^n$.

Perhaps the most useful fact about the gradient of a scalar field ϕ at a point a in $\mathbb{R}^n$ is that it is the vector at a which points in the direction of the most rapid increase of ϕ, a fact to be proved in Proposition 6.3.7.

Definition 6.3.3. Let f be a member of $\mathcal{F}_a(V, W)$ and v be a nonzero vector in V. Then $D_v f(a)$, the DERIVATIVE OF f AT a IN THE DIRECTION OF v, is defined by

$$D_v f(a) := \lim_{t \to 0} \frac{1}{t}\Delta f_a(tv)$$

if this limit exists. This directional derivative is also called the GÂTEAUX DIFFERENTIAL (or GÂTEAUX VARIATION) of f, and is sometimes denoted by $\delta f(a; v)$. Many authors require that in the preceding definition v be a unit vector. We will *not* adopt this convention.

Recall that for $\mathbf{0} \neq v \in V$ the curve $\ell \colon \mathbb{R} \to V$ defined by $\ell(t) = a + tv$ is the parametrized line through a in the direction of v. In the following proposition, which helps illuminate our use of the adjective "directional", we understand the domain of $f \circ \ell$ to be the set of all numbers t for which the expression $f(\ell(t))$ makes sense; that is,

$$\operatorname{dom}(f \circ \ell) = \{t \in \mathbb{R} \colon a + tv \in \operatorname{dom} f\}.$$

Since a is an interior point of the domain of f, the domain of $f \circ \ell$ contains an open interval about 0.

Proposition 6.3.4. *If $f \in \mathcal{D}_a(V, W)$ and $\mathbf{0} \neq v \in V$, then the directional derivative $D_v f(a)$ exists and is the tangent vector to the curve $f \circ \ell$ at 0 (where ℓ is the parametrized line through a in the direction of v). That is,*

$$D_v f(a) = D(f \circ \ell)(0).$$

Example 6.3.5. Let $f(x, y) = \ln(x^2 + y^2)^{\frac{1}{2}}$. Then $D_v f(a) = \frac{7}{10}$ when $a = (1, 1)$ and $v = \left(\frac{3}{5}, \frac{4}{5}\right)$.

Proposition 6.3.6. If $f \in \mathcal{D}_a(V, W)$, then for every nonzero v in V

$$D_v f(a) = df_a(v).$$

Proposition 6.3.7. Let $\phi \colon U \to \mathbb{R}$ be a scalar field on a subset U of $\mathbb{R}^n$. If ϕ is differentiable at a point a in U and $d\phi_a$ is not the zero functional, then the maximum value of the directional derivative $D_u \phi(a)$, taken over all unit vectors u in $\mathbb{R}^n$, is achieved when u points in the direction.of the gradient $\nabla \phi(a)$. The minimum value is achieved when u points in the opposite direction $-\nabla \phi(a)$.

What role do partial derivatives play in all this? Conceptually, not much of one. They are just directional derivatives in the directions of the standard basis vector of $\mathbb{R}^n$. They are, however, useful for computation. For example, if F is a mapping from $\mathbb{R}^n$ to $\mathbb{R}^m$ differentiable at a point a, then the matrix representation of dF_a is an $m \times n$ matrix (the so-called *Jacobian matrix*) whose entry in the jth row and k^{th} column is the partial derivative $F_k^j = \dfrac{\partial F^j}{\partial x_k}$ (where F^j is the jth coordinate function of F). And if ϕ is a differentiable scalar field on $\mathbb{R}^n$, then its gradient can be represented as $\left(\dfrac{\partial \phi}{\partial x_1}, \ldots, \dfrac{\partial \phi}{\partial x_n}\right)$.

Exercise 6.3.8. Take any elementary calculus text and derive every item called a *chain rule* in that text from Theorem 6.2.9.

Chapter 7

MULTILINEAR MAPS AND DETERMINANTS

7.1. Permutations

A bijective map $\sigma\colon X \to X$ from a set X onto itself is a PERMUTATION of the set. If $x_1, x_2, \ldots, x_n$ are distinct elements of a set X, then the permutation of X that maps $x_1 \mapsto x_2, x_2 \mapsto x_3, \ldots, x_{n-1} \mapsto x_n, x_n \mapsto x_1$ and leaves all other elements of X fixed is a CYCLE (or CYCLIC PERMUTATION) of LENGTH n. A cycle of length 2 is a TRANSPOSITION. Permutations $\sigma_1, \ldots, \sigma_n$ of a set X are DISJOINT if each $x \in X$ is moved by at most one σ_j; that is, if $\sigma_j(x) \neq x$ for at most one $j \in \mathbb{N}_n := \{1, 2, \ldots, n\}$.

Proposition 7.1.1. *If X is a nonempty set, the set of permutations of X is a group under composition.*

Notice that if σ and τ are disjoint permutations of a set X, then $\sigma\tau = \tau\sigma$. If X is a set with n elements, then the group of permutations of X (which we may identify with the group of permutations of the set $\mathbb{N}_n$) is the SYMMETRIC GROUP ON n ELEMENTS (or ON n LETTERS); it is denoted by S_n.

Proposition 7.1.2. *Any permutation $\sigma \neq \mathrm{id}_X$ of a finite set X can be written as a product (composite) of cycles of length at least 2.*

Proof. See [28], Chapter 8, Theorem 1.

A permutation of a finite set X is EVEN if it can be written as the product of an even number of transpositions, and it is ODD if it can be written as a product of an odd number of transpositions.

Proposition 7.1.3. *Every permutation of a finite set is either even or odd, but not both.*

Proof. See [28], Chapter 8, Theorem 3.

Definition 7.1.4. The SIGN of a permutation σ, denoted by $\operatorname{sgn}\sigma$, is $+1$ if σ is even and -1 if σ is odd.

7.2. Multilinear Maps

Definition 7.2.1. Let $V_1, V_2, \ldots, V_n$, and W be vector spaces over a field $\mathbb{F}$. We say that a function $f \colon V_1 \times \cdots \times V_n \to W$ is MULTILIN-EAR (or n-LINEAR) if it is linear in each of its n variables. We ordinarily call 2-linear maps BILINEAR and 3-linear maps TRILINEAR. We denote by $\mathfrak{L}^n(V_1, \ldots, V_n; W)$ the family of all n-linear maps from $V_1 \times \cdots \times V_n$ into W. A multilinear map from the product $V_1 \times \cdots \times V_n$ into the scalar field $\mathbb{F}$ is a MULTILINEAR FORM (or a MULTILINEAR FUNCTIONAL).

Exercise 7.2.2. Let V and W be vector spaces over a field $\mathbb{F}$; u, v, x, $y \in V$; and $\alpha \in \mathbb{F}$.

(a) Expand $T(u+v, x+y)$ if T is a bilinear map from $V \times V$ into W.
(b) Expand $T(u+v, x+y)$ if T is a linear map from $V \oplus V$ into W.
(c) Write $T(\alpha x, \alpha y)$ in terms of α and $T(x, y)$ if T is a bilinear map from $V \times V$ into W.
(d) Write $T(\alpha x, \alpha y)$ in terms of α and $T(x, y)$ if T is a linear map from $V \oplus V$ into W.

Example 7.2.3. Composition of operators on a vector space V is a bilinear map on $\mathfrak{L}(V)$.

Proposition 7.2.4. *If U, V, and W are vector spaces over a field $\mathbb{F}$, then so is $\mathfrak{L}^2(U, V; W)$. Furthermore the spaces $\mathfrak{L}(U, \mathfrak{L}(V, W))$ and $\mathfrak{L}^2(U, V; W)$ are (naturally) isomorphic.*

Hint for proof. The isomorphism is implemented by the map

$$F\colon \mathfrak{L}(U, \mathfrak{L}(V, W)) \to \mathfrak{L}^2(U, V; W)\colon \phi \mapsto \hat{\phi}$$

where $\hat{\phi}(u, v) := (\phi(u))(v)$ for all $u \in U$ and $v \in V$.

Definition 7.2.5. A multilinear map $f\colon V^n \to W$ from the n-fold product $V \times \cdots \times V$ of a vector space V into a vector space W is ALTERNATING if $f(v_1, \ldots, v_n) = 0$ whenever $v_i = v_j$ for some $i \neq j$.

Exercise 7.2.6. Let $V = \mathbb{R}^2$ and $f\colon V^2 \to \mathbb{R}\colon (v, w) \mapsto v_1 w_2$. Is f bilinear? Is it alternating?

Exercise 7.2.7. Let $V = \mathbb{R}^2$ and $g\colon V^2 \to \mathbb{R}\colon (v, w) \mapsto v_1 + w_2$. Is g bilinear? Is it alternating?

Exercise 7.2.8. Let $V = \mathbb{R}^2$ and $h\colon V^2 \to \mathbb{R}\colon (v, w) \mapsto v_1 w_2 - v_2 w_1$. Is h bilinear? Is it alternating? If $\{e^1, e^2\}$ is the usual basis for $\mathbb{R}^2$, what is $h(e^1, e^2)$?

Definition 7.2.9. If V and W are vector spaces, a multilinear map $f\colon V^n \to W$ is SKEW-SYMMETRIC if

$$f(v^1, \ldots, v^n) = (\operatorname{sgn} \sigma) f\big(v^{\sigma(1)}, \ldots, v^{\sigma(n)}\big) \qquad \text{for all } \sigma \in S_n.$$

Proposition 7.2.10. *Suppose that V and W be vector spaces. Then every alternating multilinear map $f\colon V^n \to W$ is skew-symmetric.*

Hint for proof. Consider $f(u + v, u + v)$ in the bilinear case.

Remark 7.2.11. If a function $f\colon \mathbb{R}^n \to \mathbb{R}$ is differentiable, then at each point a in $\mathbb{R}^n$ the differential of f at a is a linear map from $\mathbb{R}^n$ into $\mathbb{R}$. Thus we regard $df\colon a \mapsto df_a$ (the DIFFERENTIAL OF f) as a map from $\mathbb{R}^n$ into $\mathfrak{L}(\mathbb{R}^n, \mathbb{R})$. It is natural to inquire whether the function df is itself differentiable. If it is, its differential at a (which we denote by $d^2 f_a$) is a linear map from $\mathbb{R}^n$ into $\mathfrak{L}(\mathbb{R}^n, \mathbb{R})$; that is

$$d^2 f_a \in \mathfrak{L}(\mathbb{R}^n, \mathfrak{L}(\mathbb{R}^n, \mathbb{R})).$$

In the same vein, since $d^2 f$ maps $\mathbb{R}^n$ into $\mathfrak{L}(\mathbb{R}^n, \mathfrak{L}(\mathbb{R}^n, \mathbb{R}))$, its differential (if it exists) belongs to $\mathfrak{L}(\mathbb{R}^n, \mathfrak{L}(\mathbb{R}^n, \mathfrak{L}(\mathbb{R}^n, \mathbb{R})))$. It is moderately unpleasant to contemplate what an element of $\mathfrak{L}(\mathbb{R}^n, \mathfrak{L}(\mathbb{R}^n, \mathbb{R}))$ or of $\mathfrak{L}(\mathbb{R}^n, \mathfrak{L}(\mathbb{R}^n, \mathfrak{L}(\mathbb{R}^n, \mathbb{R})))$ might "look like". And clearly as we pass to even higher order differentials things look worse and worse. It is comforting to discover that an element of $\mathfrak{L}(\mathbb{R}^n, \mathfrak{L}(\mathbb{R}^n, \mathbb{R}))$ may be regarded as a map from $(\mathbb{R}^n)^2$ into $\mathbb{R}$ which is bilinear (that is, linear in both of its variables), and that $\mathfrak{L}(\mathbb{R}^n, \mathfrak{L}(\mathbb{R}^n, \mathfrak{L}(\mathbb{R}^n, \mathbb{R})))$ may be thought of as a map from $(\mathbb{R}^n)^3$ into $\mathbb{R}$ which is linear in each of its three variables. More generally, if V_1, V_2, V_3, and W are arbitrary vector spaces it will be possible to identify the vector space $\mathfrak{L}(V_1, \mathfrak{L}(V_2, W))$ with the space of bilinear maps from $V_1 \times V_2$ to W, the vector space $\mathfrak{L}(V_1, \mathfrak{L}(V_2, \mathfrak{L}(V_3, W)))$ with the trilinear maps from $V_1 \times V_2 \times V_3$ to W, and so on (see, for example, Proposition 7.2.4).

7.3. Determinants

Definition 7.3.1. A field $\mathbb{F}$ is OF CHARACTERISTIC ZERO if $n1 = 0$ for *no* $n \in \mathbb{N}$.

Convention 7.3.2. In the following material on determinants, we will assume that the scalar fields underlying all the vector spaces and algebras we encounter are of characteristic zero. Thus multilinear functions will be alternating if and only if they are skew-symmetric. (See Exercises 7.2.10 and 7.3.7.)

Remark 7.3.3. Let A be a unital commutative algebra. In the sequel we identify the algebra $(A^n)^n = A^n \times \cdots \times A^n$ (n factors) with the algebra $\mathbf{M}_n(A)$ of $n \times n$ matrices of elements of A by regarding the term a^k in $(a^1, \ldots, a^n) \in (A^n)^n$ as the kth column vector of an $n \times n$ matrix of elements of A. There are many standard notations for the same thing: $\mathbf{M}_n(A)$, $A^n \times \cdots \times A^n$ (n factors), $(A^n)^n$, $A^{n \times n}$, and A^{n^2}, for example.

The identity matrix, which we usually denote by I, in $\mathbf{M}_n(A)$ is $(e^1, \ldots, e^n)$, where $e^1, \ldots, e^n$ are the standard basis vectors for A^n; that is, $e^1 = (\mathbf{1}_A, 0, 0, \ldots)$, $e^2 = (0, \mathbf{1}_A, 0, 0, \ldots)$, and so on.

Definition 7.3.4. Let A be a unital commutative algebra. A DETERMINANT FUNCTION is an alternating multilinear map $D \colon \mathbf{M}_n(A) \to A$ such that $D(I) = \mathbf{1}_A$.

Proposition 7.3.5. *Let* $V = \mathbb{R}^n$. *Define*

$$\Delta \colon V^n \to \mathbb{R} \colon (v^1, \ldots, v^n) \mapsto \sum_{\sigma \in S_n} (\operatorname{sgn} \sigma) v^1_{\sigma(1)} \cdots v^n_{\sigma(n)}.$$

Then Δ *is a determinant function that satisfies* $\Delta(e^1, \ldots, e^n) = 1$.

Note: If A is an $n \times n$ matrix of real numbers we define $\det A$, the DETERMINANT of A, to be $\Delta(v^1, \ldots, v^n)$ where $v^1, \ldots, v^n$ are the column vectors of the matrix A.

Exercise 7.3.6. Let $A = \begin{bmatrix} 1 & 3 & 2 \\ -1 & 0 & 3 \\ -2 & -2 & 1 \end{bmatrix}$. Use the definition above to find $\det A$.

Proposition 7.3.7. *If V and W are vector spaces over a field $\mathbb{F}$ of characteristic zero and $f: V^n \to W$ is a skew-symmetric multilinear map, then f is alternating.*

Proposition 7.3.8. *Let V and W be vector spaces (over a field of characteristic 0) and $f: V^n \to W$ be a multilinear map. If $f(v^1, \ldots, v^n) = 0$ whenever two consecutive terms in the n-tuple $(v^1, \ldots, v^n)$ are equal, then f is skew-symmetric and therefore alternating.*

Proposition 7.3.9. *Let $f: V^n \to W$ be an alternating multilinear map, $j \neq k$ in $\mathbb{N}_n$, and α be a scalar. Then*

$$f(v^1, \ldots, \underset{\underset{j}{\uparrow}}{v^j + \alpha v^k}, \ldots, v^n) = f(v^1, \ldots, \underset{\underset{j}{\uparrow}}{v^j}, \ldots, v^n).$$

Proposition 7.3.10. *Let A be a unital commutative algebra (over a field of characteristic zero) and $n \in \mathbb{N}$. A determinant function exists on $\mathbf{M}_n(A)$.* Hint. *Consider*

$$\det: \mathbf{M}_n(A) \to A: (a^1, \ldots, a^n) \mapsto \sum_{\sigma \in S_n} (\operatorname{sgn} \sigma) a^1_{\sigma(1)} \cdots a^n_{\sigma(n)}.$$

Proposition 7.3.11. *Let D be an alternating multilinear map on $\mathbf{M}_n(A)$ where A is a unital commutative algebra and $n \in \mathbb{N}$. For every $C \in \mathbf{M}_n(A)$*

$$D(C) = D(I) \det C.$$

Proposition 7.3.12. *Show that the determinant function on $\mathbf{M}_n(A)$, where A is a unital commutative algebra, is unique.*

Proposition 7.3.13. *Let A be a unital commutative algebra and $B, C \in \mathbf{M}_n(A)$. Then*

$$\det(BC) = \det B \det C.$$

Hint for proof. Consider the function $D(C) = D(c^1, \ldots, c^n) := \det(Bc^1, \ldots, Bc^n)$, where Bc^k is the product of the $n \times n$ matrix B and the kth column vector of C.

Proposition 7.3.14. *For an $n \times n$ matrix B let B^t, the* TRANSPOSE *of B, be the matrix obtained from B by interchanging its rows and columns; that is, if $B = \left[b_i^j\right]$, then $B^t = \left[b_j^i\right]$. Then $\det B^t = \det B$.*

7.4. Tensor Products of Vector Spaces

For a modern and very careful exposition of tensor products, which is more extensive than given here, I recommend Chapter 14 of [29].

Definition 7.4.1. Let U and V be vector spaces over a field $\mathbb{F}$. A vector space $U \otimes V$ together with a bilinear map $\tau\colon U \times V \to U \otimes V$ is a TENSOR PRODUCT of U and V if for every vector space W and every bilinear map $B\colon U \times V \to W$, there exists a unique linear map $\widetilde{B}\colon U \otimes V \to W$ that makes the following diagram commute.

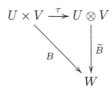

Proposition 7.4.2. *In the category of vector spaces and linear maps if tensor products exist, then they are unique (up to isomorphism).*

Proposition 7.4.3. *In the category of vector spaces and linear maps tensor products exist.*

Hint for proof. Let U and V be vector spaces over a field $\mathbb{F}$. Consider the free vector space $l_c(U \times V) = l_c(U \times V, \mathbb{F})$. Define

$$*\colon U \times V \to l_c(U \times V)\colon (u, v) \mapsto \chi_{\{(u,v)\}}.$$

Write $u * v$ instead of $*(u, v)$. Then let

$$S_1 = \{(u_1 + u_2) * v - u_1 * v - u_2 * v \colon u_1,\, u_2 \in U \text{ and } v \in V\},$$
$$S_2 = \{(\alpha u) * v - \alpha(u * v) \colon \alpha \in \mathbb{F},\, u \in U, \text{ and } v \in V\},$$
$$S_3 = \{u * (v_1 + v_2) - u * v_1 - u * v_2 \colon u \in U \text{ and } v_1,\, v_2 \in V\},$$

$$S_4 = \{u * (\alpha v) - \alpha(u * v) \colon \alpha \in \mathbb{F},\ u \in U,\ \text{and}\ v \in V\},$$
$$S = \operatorname{span}(S_1 \cup S_2 \cup S_3 \cup S_4),\ \text{and}$$
$$U \otimes V = l_c(U \times V)/S.$$

Also define

$$\tau \colon U \times V \to U \otimes V \colon (u, v) \mapsto [u * v].$$

Then show that $U \otimes V$ and τ satisfy the conditions stated in Definition 7.4.1.

Notation 7.4.4. It is conventional to write $u \otimes v$ for $\tau\big((u, v)\big) = [u * v]$. Tensors of the form $u \otimes v$ are called ELEMENTARY TENSORS (or DECOMPOSABLE TENSORS or HOMOGENEOUS TENSORS).

Proposition 7.4.5. *Let u and v be elements of finite dimensional vector spaces U and V, respectively. If $u \otimes v = \mathbf{0}$, then either $u = \mathbf{0}$ or $v = \mathbf{0}$.*

Hint for proof. Argue by contradiction. Suppose there exist $u' \neq \mathbf{0}$ in U and $v' \neq \mathbf{0}$ in V such that $u' \otimes v' = \mathbf{0}$. Use Proposition 2.5.8 to choose linear functionals $f \in U^*$ and $g \in V^*$ such that $f(u') = g(v') = 1$. Consider the map B defined on $U \times V$ by $B(u, v) = f(u)g(v)$.

Caution 7.4.6. One needs to exercise some care in dealing with elementary tensors: keep in mind that

(1) not every member of $U \otimes V$ is of the form $u \otimes v$;
(2) the representation of a tensor as an elementary tensor, even when it is possible, fails to be unique; and
(3) the family of elementary tensors (although it spans $U \otimes V$) is by no means linearly independent.

We do, however, have the following useful result.

Proposition 7.4.7. *Let $e^1, \dots, e^n$ be linearly independent vectors in a vector space U and $v^1, \dots, v^n$ be arbitrary vectors in a vector space V. Then $\sum_{j=1}^n e^j \otimes v^j = \mathbf{0}$ if and only if $v^k = \mathbf{0}$ for each $k \in \mathbb{N}_n$.*

Hint for proof. Extend $\{e^1, \dots, e^n\}$ to a basis E for U. Fix $k \in \mathbb{N}_n$. Let $\phi \in V^*$. Consider the scalar valued function B defined on $U \times V$ by $B(u, v) = \big(e^k\big)^*(u)\phi(v)$, where $\big(e^k\big)^*$ is as defined in Proposition 2.5.4. Prove that $\phi(v^k) = 0$.

Proposition 7.4.8. *If $\{e^i\}_{i=1}^m$ and $\{f^j\}_{j=1}^n$ are bases for the finite dimensional vector spaces U and V, respectively, then the family $\{e^i \otimes f^j\}_{i=1,\ j=1}^{m,\ n}$ is a basis for $U \otimes V$.*

Corollary 7.4.9. *If U and V are finite dimensional vector spaces, then so is $U \otimes V$ and*

$$\dim(U \otimes V) = (\dim U)(\dim V).$$

Proposition 7.4.10. *Let U and V be finite dimensional vector spaces and $\{f^j\}_{j=1}^n$ be a basis for V. Then for every element $t \in U \otimes V$ there exist unique vectors $u^1, \ldots, u^n \in U$ such that*

$$t = \sum_{j=1}^n u^j \otimes f^j.$$

Proposition 7.4.11. *If U and V are vector spaces, then*

$$U \otimes V \cong V \otimes U.$$

Proposition 7.4.12. *If V is a vector space over a field $\mathbb{F}$, then*

$$V \otimes \mathbb{F} \cong V \cong \mathbb{F} \otimes V.$$

Proposition 7.4.13. *Let U, V, and W be vector spaces. For every vector space X and every trilinear map $k\colon U \times V \times W \to X$ there exists a unique linear map $\widetilde{k}\colon U \otimes (V \otimes W)\colon \to X$ such that*

$$\widetilde{k}\bigl(u \otimes (v \otimes w)\bigr) = k(u, v, w)$$

for all $u \in U$, $v \in V$, and $w \in W$.

Proposition 7.4.14. *If U, V, and W are vector spaces, then*

$$U \otimes (V \otimes W) \cong (U \otimes V) \otimes W.$$

Proposition 7.4.15. *Let U and V be vector spaces. Then there exists an injective linear map from $U \otimes V^*$ into $\mathfrak{L}(U, V)$. If U and V are finite dimensional, then $U \otimes V^* \cong \mathfrak{L}(V, U)$.*

Hint for proof. Consider the map $B\colon U \times V^* \to \mathfrak{L}(V, U)\colon (u, g) \mapsto B(u, g)$ where

$$B(u, g)(v) = g(v)\, u$$

for every $v \in V$.

Proposition 7.4.16. *If U, V, and W are vector spaces, then*

$$U \otimes (V \oplus W) \cong (U \otimes V) \oplus (U \otimes W).$$

Proposition 7.4.17. *If U and V are finite dimensional vector spaces, then*

$$(U \otimes V)^* \cong U^* \otimes V^*.$$

Proposition 7.4.18. *If U, V, and W are finite dimensional vector spaces, then*

$$\mathcal{L}(U \otimes V, W) \cong \mathcal{L}(U, \mathcal{L}(V, W)) \cong \mathcal{L}^2(U, V; W).$$

Proposition 7.4.19. *Let u^1, $u^2 \in U$ and v^1, $v^2 \in V$ where U and V are finite dimensional vector spaces. If $u^1 \otimes v^1 = u^2 \otimes v^2 \neq 0$, then $u^2 = \alpha u^1$ and $v^2 = \beta v^1$ where $\alpha \beta = 1$.*

7.5. Tensor Products of Linear Maps

Definition 7.5.1. Let $S \colon U \to W$ and $T \colon V \to X$ be linear maps between vector spaces. We define the TENSOR PRODUCT of the linear maps S and T by

$$S \otimes T \colon U \otimes V \to W \otimes X \colon u \otimes v \mapsto S(u) \otimes T(v).$$

Exercise 7.5.2. Definition 7.5.1 defines the tensor product $S \otimes T$ of two maps only for homogeneous elements of $U \otimes V$. Explain exactly what is needed to convince ourselves that $S \otimes T$ is well defined on all of $U \otimes V$. Then prove that $S \otimes T$ is a linear map.

Exercise 7.5.3. Some authors hesitate to use the notation $S \otimes T$ for the mapping defined in 7.5.1 on the (very reasonable) grounds that $S \otimes T$ already has a meaning; it is a member of the vector space $\mathcal{L}(U, W) \otimes \mathcal{L}(V, X)$. Discuss this problem and explain, in particular, why the use of the notation $S \otimes T$ in 7.5.1 is not altogether unreasonable.

Proposition 7.5.4. *Suppose that R, $S \in \mathcal{L}(U, W)$ and that $T \in \mathcal{L}(V, X)$ where U, V, W, and X are finite dimensional vector spaces. Then*

$$(R + S) \otimes T = R \otimes T + S \otimes T.$$

Proposition 7.5.5. *Suppose that $R \in \mathcal{L}(U, W)$ and that S, $T \in \mathcal{L}(V, X)$ where U, V, W, and X are finite dimensional vector spaces. Then*

$$R \otimes (S + T) = R \otimes S + R \otimes T.$$

Proposition 7.5.6. *Suppose that* $S \in \mathfrak{L}(U, W)$ *and that* $T \in \mathfrak{L}(V, X)$ *where* U, V, W, *and* X *are finite dimensional vector spaces. Then for all scalars* α *and* β

$$(\alpha S) \otimes (\beta T) = \alpha\beta(S \otimes T).$$

Proposition 7.5.7. *Suppose that* $Q \in \mathfrak{L}(U, W)$, $R \in \mathfrak{L}(V, X)$, $S \in \mathfrak{L}(W, Y)$, *and that* $T \in \mathfrak{L}(X, Z)$ *where* U, V, W, X, Y, *and* Z *are finite dimensional vector spaces. Then*

$$(S \otimes T)(Q \otimes R) = SQ \otimes TR.$$

Proposition 7.5.8. *If* U *and* V *are finite dimensional vector spaces, then*

$$I_U \otimes I_V = I_{U \otimes V}.$$

Proposition 7.5.9. *Suppose that* $S \in \mathfrak{L}(U, W)$ *and that* $T \in \mathfrak{L}(V, X)$ *where* U, V, W, *and* X *are finite dimensional vector spaces. If* S *and* T *are invertible, then so is* $S \otimes T$ *and*

$$(S \otimes T)^{-1} = S^{-1} \otimes T^{-1}.$$

Proposition 7.5.10. *Suppose that* $S \in \mathfrak{L}(U, W)$ *and that* $T \in \mathfrak{L}(V, X)$ *where* U, V, W, *and* X *are finite dimensional vector spaces. If* $S \otimes T = 0$, *then either* $S = 0$ *or* $T = 0$.

Proposition 7.5.11. *Suppose that* $S \in \mathfrak{L}(U, W)$ *and that* $T \in \mathfrak{L}(V, X)$ *where* U, V, W, *and* X *are finite dimensional vector spaces. Then*

$$\operatorname{ran}(S \otimes T) = \operatorname{ran} S \otimes \operatorname{ran} T.$$

Proposition 7.5.12. *Suppose that* $S \in \mathfrak{L}(U, W)$ *and that* $T \in \mathfrak{L}(V, X)$ *where* U, V, W, *and* X *are finite dimensional vector spaces. Then*

$$\ker(S \otimes T) = \ker S \otimes V + U \otimes \ker T.$$

Chapter 8

TENSOR ALGEBRAS

8.1. Grassmann Algebras

Definition 8.1.1. Let V be an d-dimensional vector space over a field $\mathbb{F}$. We say that $\bigwedge(V)$ is the GRASSMANN ALGEBRA (or the EXTERIOR ALGEBRA) over V if

(1) $\bigwedge(V)$ is a unital algebra over $\mathbb{F}$ (multiplication is denoted by $\wedge$),
(2) V is "contained in" $\bigwedge(V)$,
(3) $v \wedge v = \mathbf{0}$ for every $v \in V$,
(4) $\dim(\bigwedge(V)) = 2^d$, and
(5) $\bigwedge(V)$ is generated by $\mathbf{1}_{\bigwedge(V)}$ and V.

The multiplication $\wedge$ in a Grassmann algebra is called the WEDGE PRODUCT (or the EXTERIOR PRODUCT).

Exercise 8.1.2. There are two instances in the preceding definition where I have opted for brevity over precision. Explain why in Definition 8.1.1 "contained in" appears in quotation marks. Give a more precise version of condition (2). Also, explain more precisely what is meant, in condition (5), by saying that $\bigwedge(V)$ is *generated by* $\mathbf{1}$ and V.

Proposition 8.1.3. *If $\bigwedge(V)$ is a Grassmann algebra over a vector space V, then the zero vector of V is an annihilator in the algebra $\bigwedge(V)$. That is, $\mathbf{0}_V \wedge g = \mathbf{0}_V$ for every $g \in \bigwedge(V)$.*

Proposition 8.1.4. *If $\bigwedge(V)$ is a Grassmann algebra over a vector space V, then $\mathbf{1}_{\bigwedge(V)} \notin V$.*

Proposition 8.1.5. *Let v and w be elements of a finite dimensional vector space V. In the Grassmann algebra $\bigwedge(V)$ generated by V*

$$v \wedge w = -w \wedge v.$$

Proposition 8.1.6. *Let V be a d-dimensional vector space with basis $E = \{e^1, \ldots, e^d\}$. For each nonempty subset $S = \{e^{i_1}, e^{i_2}, \ldots, e^{i_p}\}$ of E with $i_1 < i_2 < \cdots < i_p$ let $e_S = e^{i_1} \wedge e^{i_2} \wedge \cdots \wedge e^{i_p}$. Also let $e_\emptyset = \mathbf{1}_{\bigwedge(V)}$. Then $\{e_S \colon S \subseteq E\}$ is a basis for the Grassmann algebra $\bigwedge(V)$.*

Definition 8.1.7. An algebra A is a $\mathbb{Z}^+$-GRADED ALGEBRA if it is a direct sum $A = \bigoplus_{k \geq 0} A_k$ of vector subspaces A_k and its multiplication $\wedge$ takes elements in $A_j \times A_k$ to elements in A_{j+k} for all j, $k \in \mathbb{Z}^+$. An element in A_k is said to be a HOMOGENEOUS element of DEGREE k (or of GRADE k).

The definitions of $\mathbb{Z}$-graded algebras, $\mathbb{N}$-graded algebras and $\mathbb{Z}_2$-graded algebras are similar. (In the case of a Z_2-graded algebra the indices are 0 and 1 and $A_1 \wedge A_1 \subseteq A_0$.) Usually the unmodified expression "graded algebra" refers to a $\mathbb{Z}^+$-graded algebra.

Proposition 8.1.8 says that every Grassmann algebra $\bigwedge(V)$ over a vector space V is a graded algebra. The set of elements homogeneous of degree k is denoted by $\bigwedge^k(V)$. An element of $\bigwedge^k(V)$ that can be written in the form $v_1 \wedge v_2 \wedge \cdots \wedge v_k$ (where $v_1, \ldots, v_k$ all belong to V) is a DECOMPOSABLE element of DEGREE k (or of GRADE k).

Proposition 8.1.8. *Every Grassmann algebra is a $\mathbb{Z}^+$-graded algebra.*

We denote by $\bigwedge^k(V)$ the subspace of all homogeneous elements of degree k in $\bigwedge(V)$. In particular, $\bigwedge^0(V) = \mathbb{F}$ and $\bigwedge^1(V) = V$. If the dimension of V is d, take $\bigwedge^k(V) = \{\mathbf{0}\}$ for all $k > d$. (And if you wish to regard $\bigwedge(V)$ as a $\mathbb{Z}$-graded algebra also take $\bigwedge^k(V) = \{\mathbf{0}\}$ whenever $k < 0$.)

Example 8.1.9. If the dimension of a vector space V is 3 or less, then every homogeneous element of the corresponding Grassmann algebra is decomposable.

Example 8.1.10. If the dimension of a (finite dimensional) vector space V is at least four, then there exist homogeneous elements in the corresponding Grassmann algebra that are not decomposable.

Hint for proof. Let e^1, e^2, e^3, and e^4 be distinct basis elements of V and consider $(e^1 \wedge e^2) + (e^3 \wedge e^4)$.

Proposition 8.1.11. *The elements $v_1, v_2, \ldots, v_p$ in a vector space V are linearly independent if and only if $v_1 \wedge v_2 \wedge \cdots \wedge v_p \neq \mathbf{0}$ in the corresponding Grassmann algebra $\bigwedge(V)$.*

Proposition 8.1.12. *Let $T \colon V \to W$ be a linear map between finite dimensional vector spaces. Then there exists a unique extension of T to a unital algebra homomorphism $\bigwedge(T) \colon \bigwedge(V) \to \bigwedge(W)$. This extension maps $\bigwedge^k(V)$ into $\bigwedge^k(W)$ for each $k \in \mathbb{N}$.*

Example 8.1.13. The pair of maps $V \mapsto \bigwedge(V)$ and $T \mapsto \bigwedge(T)$ is a covariant functor from the category of vector spaces and linear maps to the category of unital algebras and unital algebra homomorphisms.

Proposition 8.1.14. *If V is a vector space of dimension d, then $\dim\bigl(\bigwedge^p(V)\bigr) = \binom{d}{p}$ for $0 \leq p \leq d$.*

Convention 8.1.15. Let V be a d-dimensional vector space. Since the map $\lambda \mapsto \lambda \mathbf{1}_{\bigwedge(V)}$ is an obvious isomorphism between $\mathbb{F}$ and the one-dimensional space $\bigwedge^0(V)$, we identify these two spaces, and refer to an element of $\bigwedge^0(V)$ as a SCALAR. And since $\bigwedge^d(V)$ is also one-dimensional, its elements are frequently referred to as PSEUDOSCALARS.

Proposition 8.1.16. *If V is a finite dimensional vector space, $\omega \in \bigwedge^p(V)$, and $\mu \in \bigwedge^q(V)$, then*

$$\omega \wedge \mu = (-1)^{pq} \mu \wedge \omega.$$

8.2. Existence of Grassmann Algebras

Definition 8.2.1. Let $V_0, V_1, V_2, \ldots$ be vector spaces (over the same field). Then their (EXTERNAL) DIRECT SUM, which is denoted by $\bigoplus_{k=0}^{\infty} V_k$, is defined to be the set of all functions $v \colon \mathbb{Z}^+ \to \bigcup_{k=0}^{\infty} V_k$ with finite support such that $v(k) = v_k \in V_k$ for each $k \in \mathbb{Z}^+$. The usual pointwise addition and scalar multiplication make this set into a vector space.

Definition 8.2.2. Let V be a vector space over a field $\mathbb{F}$. Define $\mathcal{T}^0(V) = \mathbb{F}$, $\mathcal{T}^1(V) = V$, $\mathcal{T}^2(V) = V \otimes V$, $\mathcal{T}^3(V) = V \otimes V \otimes V, \ldots, \mathcal{T}^k(V) = V \otimes \cdots \otimes V$ (k factors), ... Then let $\mathcal{T}(V) = \bigoplus_{k=0}^{\infty} \mathcal{T}^k(V)$. Define multiplication on $\mathcal{T}(V)$ by using the obvious isomorphism

$$\mathcal{T}^k(V) \otimes \mathcal{T}^m(V) \cong \mathcal{T}^{k+m}(V)$$

and extending by linearity to all of $\mathcal{T}(V)$. The resulting algebra is the TENSOR ALGEBRA of V (or generated by V).

Proposition 8.2.3. *The tensor algebra $\mathcal{T}(V)$ as defined in 8.2.2 is in fact a unital algebra.*

Proposition 8.2.4. *Let V be a finite dimensional vector space and J be the ideal in the tensor algebra $\mathcal{T}(V)$ generated by the set of all elements of the form $v \otimes v$ where $v \in V$. Then the quotient algebra $\mathcal{T}(V)/J$ is the Grassmann algebra over V^* (or, equivalently, over V).*

Notation 8.2.5. If x and y are elements of the tensor algebra $\mathcal{T}(V)$, then in the quotient algebra $\mathcal{T}(V)/J$ the product of $[x]$ and $[y]$ is written using "wedge notation"; that is,

$$[x] \wedge [y] = [x \otimes y].$$

Notation 8.2.6. If V is a vector space over $\mathbb{F}$ and $k \in \mathbb{N}$ we denote by $\mathrm{Alt}^k(V)$ the set of all alternating k-linear maps from V^k into $\mathbb{F}$. (The space $\mathrm{Alt}^1(V)$ is just V^*.) Additionally, take $\mathrm{Alt}^0(V) = \mathbb{F}$.

Example 8.2.7. If V is a finite dimensional vector space and $k > \dim V$, then $\mathrm{Alt}^k(V) = \{\mathbf{0}\}$.

Definition 8.2.8. Let $p, q \in \mathbb{N}$. We say that a permutation $\sigma \in S_{p+q}$ is a (p, q)-SHUFFLE if $\sigma(1) < \cdots < \sigma(p)$ and $\sigma(p+1) < \cdots < \sigma(p+q)$. The set of all such permutations is denoted by $S(p, q)$.

Example 8.2.9. Give an example of a $(4,5)$-shuffle permutation σ of the set $\mathbb{N}_9 = \{1, \ldots, 9\}$ such that $\sigma(7) = 4$.

Definition 8.2.10. Let V be a vector space over a field of characteristic 0. For $p, q \in \mathbb{N}$ define

$$\wedge \colon \mathrm{Alt}^p(V) \times \mathrm{Alt}^q(V) \to \mathrm{Alt}^{p+q}(V) \colon (\omega, \mu) \mapsto \omega \wedge \mu$$

where

$$(\omega \wedge \mu)(v^1, \ldots, v^{p+q})$$
$$= \sum_{\sigma \in S(p,q)} (\mathrm{sgn}\,\sigma)\omega(v^{\sigma(1)}, \ldots, v^{\sigma(p)})\mu(v^{\sigma(p+1)}, \ldots, v^{\sigma(p+q)}).$$

Exercise 8.2.11. Show that Definition 8.2.10 is not overly optimistic by verifying that if $\omega \in \text{Alt}^p(V)$ and $\mu \in \text{Alt}^q(V)$, then $\omega \wedge \mu \in \text{Alt}^{p+q}(V)$.

Proposition 8.2.12. *The multiplication defined in 8.2.10 is associative. That is if $\omega \in \text{Alt}^p(V)$, $\mu \in \text{Alt}^q(V)$, and $\nu \in \text{Alt}^r(V)$, then*

$$\omega \wedge (\mu \wedge \nu) = (\omega \wedge \mu) \wedge \nu.$$

Exercise 8.2.13. Let V be a finite dimensional vector space over a field of characteristic zero. Explain in detail how to make $\text{Alt}^k(V)$ (or, if you prefer, $\text{Alt}^k(V^*)$) into a vector space for each $k \in \mathbb{Z}$ and how to make the collection of these into a $\mathbb{Z}$-graded algebra. Show that this algebra is the Grassmann algebra generated by V. *Hint.* Take $\text{Alt}^k(V) = \{0\}$ for each $k < 0$ and extend the definition of the wedge product so that if $\alpha \in \text{Alt}^0(V) = \mathbb{F}$ and $\omega \in \text{Alt}^p(V)$, then $\alpha \wedge \omega = \alpha\omega$.

Proposition 8.2.14. *Let $\omega_1, \ldots, \omega_p$ be members of $\text{Alt}^1(V)$ (that is, linear functionals on V). Then*

$$(\omega_1 \wedge \cdots \wedge \omega_p)(v^1, \ldots, v^p) = \det\left[\omega_j(v^k)\right]_{j,k=1}^p$$

for all $v^1, \ldots, v^p \in V$.

Proposition 8.2.15. *If $\{e_1, \ldots, e_n\}$ is a basis for an n-dimensional vector space V, then*

$$\{e^*_{\sigma(1)} \wedge \cdots \wedge e^*_{\sigma(p)} : \sigma \in S(p, n-p)\}$$

is a basis for $\text{Alt}^p(V)$.

Proposition 8.2.16. *For $T \colon V \to W$ a linear map between vector spaces define*

$$\text{Alt}^p(T) \colon \text{Alt}^p(W) \to \text{Alt}^p(V) \colon \omega \mapsto \text{Alt}^p(T)(\omega)$$

where $\left[\text{Alt}^p(T)(\omega)\right](v^1, \ldots, v^p) = \omega(Tv^1, \ldots, Tv^p)$ for all $v^1, \ldots, v^p \in V$. Then Alt^p is a contravariant functor from the category of vector spaces and linear maps into itself.

Exercise 8.2.17. Let V be an n-dimensional vector space and $T \in \mathfrak{L}(V)$. If T is diagonalizable, then

$$c_T(\lambda) = \sum_{k=0}^n (-1)^k [\text{Alt}^{n-k}(T)] \lambda^k.$$

8.3. The Hodge *-operator

Definition 8.3.1. Let E be a basis for an n-dimensional vector space V. Then the n-tuple $(e^1, \ldots, e^n)$ is an ORDERED BASIS for V if $e^1, \ldots, e^n$ are the distinct elements of E.

Definition 8.3.2. Let $E = (e^1, \ldots, e^n)$ be an ordered basis for $\mathbb{R}^n$. We say that the basis E is RIGHT-HANDED if $\det[e^1, \ldots, e^n] > 0$ and LEFT-HANDED otherwise.

Definition 8.3.3. Let V be a real n-dimensional vector space and $T \colon \mathbb{R}^n \to V$ be an isomorphism. Then the set of all n-tuples of the form $(T(e^1), \ldots, T(e^n))$ where $(e^1, \ldots, e^n)$ is a right-handed basis in $\mathbb{R}^n$ is an ORIENTATION of V. Another orientation consists of the set of n-tuples $(T(e^1), \ldots, T(e^n))$ where $(e^1, \ldots, e^n)$ is a left-handed basis in $\mathbb{R}^n$. Each of these orientations is the OPPOSITE (or REVERSE) of the other. A vector space together with one of these orientations is an ORIENTED VECTOR SPACE.

Exercise 8.3.4. Let V be an n-dimensional real inner product space. In Exercise 2.5.14 we established an isomorphism $\Phi \colon v \mapsto v^*$ between V and its dual space V^*. Show how this isomorphism can be used to induce an inner product on V^*. Then show how this may be used to create an inner product on $\operatorname{Alt}^p(V)$ for $2 \le p \le n$. *Hint.* For $v, w \in V$ let $\langle v^*, w^* \rangle = \langle v, w \rangle$. Then for $\omega_1, \ldots, \omega_p, \mu_1, \ldots, \mu_p \in \operatorname{Alt}^1(V)$ let $\langle \omega_1 \wedge \cdots \wedge \omega_p, \mu_1 \wedge \cdots \wedge \mu_p \rangle = \det[\langle \omega_j, \mu_k \rangle]$.

Definition 8.3.5. Let V be an n-dimensional oriented real inner product space. Fix a unit vector $\operatorname{vol} \in \operatorname{Alt}^d(V)$. This vector is called a VOLUME ELEMENT. (In the case where $V = \mathbb{R}^n$, we will always choose $\operatorname{vol} = e_1^* \wedge \cdots \wedge e_n^*$ where $(e_1, \ldots, e_n)$ is the usual ordered basis for $\mathbb{R}^n$.)

Proposition 8.3.6. *If V is an n-dimensional oriented real inner product space, $\omega \in \operatorname{Alt}^p(V)$, and $q = n - p$, then there exists a vector $*\omega \in \operatorname{Alt}^q(V)$ such that*

$$\langle *\omega, \mu \rangle \operatorname{vol} = \omega \wedge \mu$$

for each $\mu \in \text{Alt}^q(V)$. *Furthermore, the map* $\omega \mapsto *\omega$ *from* $\text{Alt}^p(V)$ *into* $\text{Alt}^q(V)$ *is a vector space isomorphism. This map is the* HODGE STAR OPERATOR.

Proposition 8.3.7. *Let V be a finite dimensional oriented real inner product space of dimension n. Suppose that $p+q = n$. Then $**\omega = (-1)^{pq}\omega$ for every $\omega \in \text{Alt}^p(V)$.*

Chapter 9

DIFFERENTIAL MANIFOLDS

The purpose of this chapter and the next is to examine an important and nontrivial example of a Grassmann algebra: the algebra of differential forms on a differentiable manifold. If your background includes a study of manifolds, skip this chapter. If you are pressed for time, reading just the first section of the chapter should enable you to make sense out of most of the ensuing material. That section deals with familiar manifolds in low (three or less) dimensional Euclidean spaces. The major weakness of this presentation is that it treats manifolds in a noncoordinate-free manner as subsets of some larger Euclidean space. (A helix, for example, is a 1-manifold embedded in 3-space.) The rest of the chapter gives a (very brief) introduction to a more satisfactory coordinate-free way of viewing manifolds. For a less sketchy view of the subject read one of the many splendid introductory books in the field. I particularly like [3] and [21].

9.1. Manifolds in $\mathbb{R}^3$

A 0-MANIFOLD is a point (or finite collection of points).

A function is SMOOTH if it is infinitely differentiable (that is, if it has derivatives of all orders).

A CURVE is a continuous image in $\mathbb{R}^3$ of a closed line segment in $\mathbb{R}$. If C is a curve, the choice of an interval $[a, b]$ and a continuous function f such that $C = f^{\rightarrow}([a, b])$ is a PARAMETRIZATION of C. If the function f is smooth, we say that C is a SMOOTH CURVE.

A 1-MANIFOLD is a curve (or finite collection of curves). A 1-manifold is FLAT if it is contained in some line in $\mathbb{R}^3$. For example, the line segment connecting two points in $\mathbb{R}^3$ is a flat 1-manifold.

A SURFACE is a continuous image of a closed rectangular region in $\mathbb{R}^2$. If S is a surface, the choice of a rectangle $R = [a_1, b_1] \times [a_2, b_2]$ and a continuous function f such that $S = f^\rightarrow(R)$ is a PARAMETRIZATION of S. If the function f is smooth, we say that S is a SMOOTH SURFACE.

A 2-MANIFOLD is a surface (or finite collection of surfaces). A 2-manifold is FLAT if it is contained in some plane in $\mathbb{R}^3$. For example, the triangular region connecting the points $(1, 0, 0)$, $(0, 1, 0)$, and $(0, 0, 1)$ is a flat 2-manifold.

A SOLID is a continuous image of the 3-dimensional region determined by a closed rectangular parallelepiped (to avoid a six-syllable word many people say *rectangular solid* or even just *box*) in $\mathbb{R}^3$. If E is a solid, then the choice of a rectangular parallelepiped $P = [a_1, b_1] \times [a_2, b_2] \times [a_3, b_3]$ and a continuous function f such that $E = f^\rightarrow(P)$ is a PARAMETRIZATION of E. If the function f is smooth, we say that E is a SMOOTH SOLID.

A 3-MANIFOLD is a solid (or finite collection of solids).

9.2. Charts, Atlases, and Manifolds

Definition 9.2.1. Let M and S be sets; U, $V \subseteq M$, $\phi\colon U \to S$, and $\psi\colon V \to S$ be injective maps. Then the COMPOSITE $\psi \circ \phi^{-1}$, is taken to be the function

$$\psi \circ \phi^{-1}\colon \phi^\rightarrow(U \cap V) \to \psi^\rightarrow(U \cap V).$$

These composite maps are called, variously, TRANSITION MAPS or OVERLAP MAPS or CONNECTING MAPS.

Proposition 9.2.2. *The preceding definition makes sense and the composite map $\psi \circ \phi^{-1}$ is a bijection.*

Definition 9.2.3. Let m, $n \in \mathbb{N}$, and $U \overset{\circ}{\subseteq} \mathbb{R}^m$. A function $F\colon U \to \mathbb{R}^n$ is SMOOTH (or INFINITELY DIFFERENTIABLE, or $\mathcal{C}^\infty$) if the differential $d^p F_a$ exists for every $p \in \mathbb{N}$ and every $a \in U$. We denote by $\mathcal{C}^\infty(U, \mathbb{R}^n)$ the family of all smooth functions from U into $\mathbb{R}^n$.

Definition 9.2.4. Let M be a topological space and $n \in \mathbb{N}$. A pair (U, ϕ), where U is an open subset of M and $\phi\colon U \to \widetilde{U}$ is a homeomorphism from

U to an open subset $\widetilde{U}$ of $\mathbb{R}^n$, is called an (n-DIMENSIONAL COORDINATE) CHART. (The notation here is a bit redundant. If we know the function ϕ, then we also know its domain U. Indeed, do not be surprised to see reference to *the chart ϕ* or to *the chart U*.)

Let $p \in M$. A chart (U, ϕ) is said to CONTAIN p if $p \in U$ and is said to be a chart (CENTERED) AT p if $\phi(p) = \mathbf{0}$. A family of n-dimensional coordinate charts whose domains cover M is an (n-DIMENSIONAL) ATLAS for M. If such an atlas exists, the space M is said to be LOCALLY EUCLIDEAN.

Notation 9.2.5. Let $n \in \mathbb{N}$. For $1 \leq k \leq n$ the function $\pi_k \colon \mathbb{R}^n \to \mathbb{R} \colon x = (x_1, x_2, \ldots, x_n) \mapsto x_k$ is the kth COORDINATE PROJECTION. If $\phi \colon M \to \mathbb{R}^n$ is a chart on a topological space, one might reasonably expect the n component functions of ϕ (that is, the functions $\pi_k \circ \phi$) to be called $\phi^1, \ldots, \phi^n$. But this is uncommon. People seem to like ϕ and ψ as names for charts. But then the components of these maps have names such as $x^1, \ldots, x^n$, or $y^1, \ldots, y^n$. Thus we usually end up with something like $\phi(p) = \big(x^1(p), \ldots, x^n(p)\big)$. The numbers $x^1(p), \ldots, x^n(p)$ are called the LOCAL COORDINATES of p.

Two common exceptions to this notational convention occur in the cases when $n = 2$ or $n = 3$. In the former case you are likely to see things like $\phi = (x, y)$ and $\psi = (u, v)$ for charts on 2-manifolds. Similarly, for 3-manifolds expect to see notations such as $\phi = (x, y, z)$ and $\psi = (u, v, w)$.

Definition 9.2.6. A second countable Hausdorff topological space M equipped with an n-dimensional atlas is a TOPOLOGICAL n-MANIFOLD (or just a TOPOLOGICAL MANIFOLD).

Definition 9.2.7. Charts ϕ and ψ of a topological n-manifold M are said to be (SMOOTHLY) COMPATIBLE if the transition maps $\psi \circ \phi^{-1}$ and $\phi \circ \psi^{-1}$ are smooth. An atlas on M is a SMOOTH ATLAS if every pair of its charts is smoothly compatible. Two atlases on M are (SMOOTHLY) COMPATIBLE (or EQUIVALENT) if every chart of one atlas is smoothly compatible with every chart of the second; that is, if their union is a smooth atlas on M.

Proposition 9.2.8. *Every smooth atlas on a topological manifold is contained in a unique maximal smooth atlas.*

Definition 9.2.9. A maximal smooth atlas on a topological manifold M is a DIFFERENTIAL STRUCTURE on M. A topological n-manifold that has been given a differential structure is a SMOOTH n-MANIFOLD (or a DIFFERENTIAL n-MANIFOLD, or a $\mathcal{C}^\infty$ n-MANIFOLD).

NOTE: From now on we will be concerned only with differential manifolds; so the modifier "smooth" will ordinarily be omitted when we refer to charts, to atlases, or to manifolds. Thus it will be understood that by *manifold* we mean a topological n-manifold (for some fixed n) equipped with a differential structure to which all the charts we mention belong.

Example 9.2.10. Let U be an open subset of $\mathbb{R}^n$ for some $n \in \mathbb{Z}^+$ and $\iota\colon U \to \mathbb{R}^n\colon x \mapsto x$ be the inclusion map. Then $\{\iota\}$ is a smooth atlas for U. We make the convention that when an open subset of $\mathbb{R}^n$ is regarded as an n-manifold we will suppose, unless the contrary is explicitly stated, that the inclusion map ι is a chart in its differentiable structure.

Example 9.2.11 (An atlas for $\mathbb{S}^1$). Let $\mathbb{S}^1 = \{(x,y) \in \mathbb{R}^2\colon x^2 + y^2 = 1\}$ be the unit circle in $\mathbb{R}^2$ and $U = \{(x,y) \in \mathbb{S}^1\colon y \neq -1\}$. Define $\phi\colon U \to \mathbb{R}$ to be the projection of points in U from the point $(0,-1)$ onto the x-axis; that is, if $p = (x,y)$ is a point in U, then $\big(\phi(p),0\big)$ is the unique point on the x-axis that is collinear with both $(0,-1)$ and (x,y).

(1) Find an explicit formula for ϕ.
(2) Let $V = \{(x,y) \in \mathbb{S}^1\colon y \neq 1\}$. Find an explicit formula for the projection ψ of points in V from $(0,1)$ onto the x-axis.
(3) The maps ϕ and ψ are bijections between U and V, respectively, and $\mathbb{R}$.
(4) The set $\{\phi,\psi\}$ is a (smooth) atlas for $\mathbb{S}^1$.

Example 9.2.12 (An atlas for the n-sphere). The previous example can be generalized to the n-sphere $\mathbb{S}^n = \{x \in \mathbb{R}^{n+1}\colon \|x\| = 1\}$. The generalization of the mapping ϕ is called the STEREOGRAPHIC PROJECTION FROM THE SOUTH POLE and the generalization of ψ is the STEREOGRAPHIC PROJECTION FROM THE NORTH POLE. (Find a simple expression for the transition maps.)

Example 9.2.13 (Another atlas for $\mathbb{S}^1$). Let $U = \{(x,y) \in \mathbb{S}^1\colon x \neq 1\}$. For $(x,y) \in U$ let $\phi(x,y)$ be the angle (measured counterclockwise) at the origin from $(1,0)$ to (x,y). (So $\phi(x,y) \in (0,2\pi)$.) Let $V = \{(x,y) \in \mathbb{S}^1\colon y \neq 1\}$. For $(x,y) \in V$ let $\psi(x,y)$ be $\pi/2$ plus the angle (measured counterclockwise) at the origin from $(0,1)$ to (x,y). (So $\psi(x,y) \in (\pi/2, 5\pi/2)$.) Then $\{\phi,\psi\}$ is a (smooth) atlas for $\mathbb{S}^1$.

Example 9.2.14 (The projective plane $\mathbb{P}^2$). Let $\mathbb{P}^2$ be the set of all lines through the origin in $\mathbb{R}^3$. Such a line is determined by a nonzero vector lying on the line. Two nonzero vectors $x = (x^1, x^2, x^3)$ and $y = (y^1, y^2, y^3)$

determine the same line if there exists $\alpha \in \mathbb{R}$, $\alpha \neq 0$, such that $y = \alpha x$. In this case we write $x \sim y$. It is clear that $\sim$ is an equivalence relation. We regard a member of $\mathbb{P}^2$ as an equivalence class of nonzero vectors. Let $U_k = \{[x] \in \mathbb{P}^2 : x^k \neq 0\}$ for $k = 1, 2, 3$. Also let

$$\phi \colon U_1 \to \mathbb{R}^2 \colon [(x,y,z)] \mapsto \left(\tfrac{y}{x}, \tfrac{z}{x}\right);$$
$$\psi \colon U_2 \to \mathbb{R}^2 \colon [(x,y,z)] \mapsto \left(\tfrac{x}{y}, \tfrac{z}{y}\right); \text{ and}$$
$$\eta \colon U_3 \to \mathbb{R}^2 \colon [(x,y,z)] \mapsto \left(\tfrac{x}{z}, \tfrac{y}{z}\right).$$

The preceding sets and maps are well defined; and $\{\phi, \psi, \eta\}$ is a (smooth) atlas for $\mathbb{P}^2$.

Example 9.2.15 (The general linear group). Let $G = GL(n, \mathbb{R})$ be the group of nonsingular $n \times n$ matrices of real numbers. If $\mathbf{a} = [a_{jk}]$ and $\mathbf{b} = [b_{jk}]$ are members of G define

$$d(\mathbf{a}, \mathbf{b}) = \left[\sum_{j,k=1}^{n} (a_{jk} - b_{jk})^2\right]^{\frac{1}{2}}.$$

The function d is a metric on G. Define

$$\phi \colon G \to \mathbb{R}^{n^2} \colon \mathbf{a} = [a_{jk}] \mapsto \left(a_{11}, \ldots, a_{1n}, a_{21}, \ldots, a_{2n}, \ldots, a_{n1}, \ldots, a_{nn}\right).$$

Then $\{\phi\}$ is a (smooth) atlas on G. (Be a little careful here. There is one point that is not completely obvious.)

Example 9.2.16. Let

$$I \colon \mathbb{R} \to \mathbb{R} \colon x \mapsto x,$$
$$a \colon \mathbb{R} \to \mathbb{R} \colon x \mapsto \arctan x, \text{ and}$$
$$c \colon \mathbb{R} \to \mathbb{R} \colon x \mapsto x^3.$$

Each of $\{I\}$, $\{a\}$, and $\{c\}$ is a (smooth) atlas for $\mathbb{R}$. Which of these are equivalent?

9.3. Differentiable Functions Between Manifolds

Definition 9.3.1. A function $F \colon M \to N$ between (smooth) manifolds is SMOOTH AT a point $m \in M$ if there exist charts (U, ϕ) containing m and (V, ψ) containing $F(m)$ such that $F^{\to}(U) \subseteq V$ and the LOCAL REPRESENTATIVE $F_{\psi\phi} := \psi \circ F \circ \phi^{-1} \colon \phi^{\to}(U) \to \psi^{\to}(V)$ is smooth at $\phi(m)$. The map F is SMOOTH if it is smooth at every $m \in M$.

In the case that N is a subset of some Euclidean space $\mathbb{R}^n$, it is the usual convention to use the inclusion mapping $\iota\colon N \to \mathbb{R}^n$ as the preferred chart on N. In this case the local representative of F is written as F_ϕ rather than $F_{\iota\phi}$.

NOTE: It now makes sense to say (and is true) that a single chart on a manifold is a smooth map.

Proposition 9.3.2. *Suppose that a map $F\colon M \to N$ between manifolds is smooth at a point m and that (W, μ) and (X, ν) are charts at m and $F(m)$, respectively, such that $F^\to(W) \subseteq X$. Then $\nu \circ F \circ \mu^{-1}$ is smooth at $\mu(m)$.*

Proposition 9.3.3. *If $F\colon M \to N$ and $G\colon N \to P$ are smooth maps between manifolds, then $G \circ F$ is smooth.*

Proposition 9.3.4. *Every smooth map $F\colon M \to N$ between manifolds is continuous.*

Example 9.3.5. Consider the 2-manifold $\mathbb{S}^2$ with the differential structure generated by the stereographic projections form the north and south poles (see Example 9.2.12) and the 2-manifold $\mathbb{P}^2$ with the differentiable structure generated by the atlas given in Example 9.2.14. The map $F\colon \mathbb{S}^2 \to \mathbb{P}^2\colon (x, y, z) \mapsto [(x, y, z)]$ is smooth. (Think of F as taking a point (x, y, z) in $\mathbb{S}^2$ to the line in $\mathbb{R}^3$ passing through this point and the origin.)

9.4. The Geometric Tangent Space

Definition 9.4.1. Let $J \subseteq \mathbb{R}$ be an interval with nonempty interior and M be a manifold. A smooth function $c\colon J \to M$ is a CURVE in M. If, in addition, $0 \in J^\circ$ and $c(0) = m \in M$, then we say that c is a CURVE AT m.

Example 9.4.2. The function $b\colon t \mapsto t^2$ is a curve at 0 in $\mathbb{R}$. Sketch the range of b.

The next example illustrates the point that the ranges of smooth curves may not "look smooth".

Example 9.4.3. The function $c\colon t \mapsto (\cos^3 t, \sin^3 t)$ is a curve at $(1, 0)$ in $\mathbb{R}^2$. Sketch its range.

Let V and W be normed linear spaces. Recall that a W-valued function f defined on some neighborhood of 0 in V is said to belong to the family

$\mathfrak{o}(V, W)$ provided that for every $\epsilon > 0$ there exists $\delta > 0$ such that $\|f(x)\| \leq \epsilon \|x\|$ whenever $\|x\| \leq \delta$. Recall also that W-valued functions f and g, each defined on a neighborhood of 0 in V, are said to be TANGENT (at 0) if $f - g \in \mathfrak{o}(V, W)$. In this case we write $f \simeq g$.

Thus in the special case when b and c are curves at a point w in a normed linear space W, we *should* say that b and c are tangent *at* 0 if $b - c \in \mathfrak{o}(\mathbb{R}, W)$. As a matter of fact, it is almost universal custom in this situation to say that b and c are tangent *at* w. This use of "at w" for "at 0" results from a tendency to picture curves in terms of their ranges. (For example, asked to visualize the curve $c \colon t \mapsto (\cos t, \sin t)$, most people see a circle. Of course, the circle is only the range of c and not its graph, which is a helix in $\mathbb{R}^3$.) We will follow convention and say that the curves b and c are tangent *at* w. This convention will also apply to curves in manifolds.

Example 9.4.4. It is important to note that it cannot be determined whether two curves are tangent just by looking at their ranges. The curves $b \colon t \mapsto (t, t^2)$ and $c \colon t \mapsto (2t, 4t^2)$ have identical ranges; they both follow the parabola $y = x^2$ in $\mathbb{R}^2$ (and in the same direction). They are both curves at the origin. Nevertheless, they are *not tangent* at the origin.

Definition 9.4.6 below says that curves in a manifold are *tangent* if their composites with a chart ϕ are tangent in the sense described above. Before formally stating this definition it is good to know that tangency thus defined does not depend on the particular chart chosen.

Proposition 9.4.5. *Let m be a point in a manifold and b and c be curves at m. If $\phi \circ b \simeq \phi \circ c$ for some chart ϕ centered at m, then $\psi \circ b \simeq \psi \circ c$ for every chart ψ centered at m.*

Hint for proof. Use Proposition 25.4.7 and Problem 25.4.10 in [12].

Definition 9.4.6. Let m be a point in a manifold and b and c be curves at m. Then b and c are TANGENT AT m (we write $b \simeq c$) if $\phi \circ b \simeq \phi \circ c$ for some (hence all) charts ϕ centered at m.

It is useful to know that smooth mappings between manifolds preserve tangency.

Proposition 9.4.7. *If $F \colon M \to N$ is a smooth mapping between manifolds and b and c are curves tangent at a point $m \in M$, then the curves $F \circ b$ and $F \circ c$ are tangent at $F(m)$ in N.*

Definition 9.4.8. Since the family of "little-oh" functions is closed under addition it is obvious that tangency at a point m is an equivalence relation on the family of curves at m. We denote the equivalence class containing the curve c by $\tilde{c}$ or, if we wish to emphasize the role of the point m, by $\tilde{c}_m$. Each equivalence class $\tilde{c}_m$ is a GEOMETRIC TANGENT VECTOR at m and the family of all such vectors is the GEOMETRIC TANGENT SPACE at m. The geometric tangent space at m is denoted by $\tilde{T}_m$ (or, if we wish to emphasize the role of the manifold M, by $\tilde{T}_m(M)$).

The language (involving the words "vector" and "space") in the preceding definition is highly optimistic. So far we have a *set* of equivalence classes — with no vector space structure. The key to providing $\tilde{T}_m$ with a such a structure is Exercise 2.2.16. There we found that a set S may be given a vector space structure by transferring the structure from a known vector space V to the set S by means of a bijection $f\colon S \to V$. We will show, in particular, that if M is an n-manifold, then for each $m \in M$ the geometric tangent space $\tilde{T}_m$ can be given the vector space structure of $\mathbb{R}^n$.

Definition 9.4.9. Let ϕ be a chart containing a point m in an n-manifold M and let $\mathbf{u}$ be a nonzero vector in $\mathbb{R}^n$. For every $t \in \mathbb{R}$ such that $\phi(m) + t\mathbf{u}$ belongs to the range of ϕ, let

$$c_{\mathbf{u}}(t) = \phi^{-1}\big(\phi(m) + t\mathbf{u}\big).$$

Notice that since $c_{\mathbf{u}}$ is the composite of smooth functions and since $c_{\mathbf{u}}(0) = m$, it is clear that $c_{\mathbf{u}}$ is a curve at m in M.

Example 9.4.10. If M is an n-manifold, then the curves $c_{\mathbf{e}^1}, \ldots, c_{\mathbf{e}^n}$ obtained by means of the preceding definition from the standard basis vectors $\mathbf{e}^1, \ldots, \mathbf{e}^n$ of $\mathbb{R}^n$ will prove to be very useful. We will shorten the notation somewhat and write $\mathbf{c}_k$ for $c_{\mathbf{e}^k}$ ($1 \le k \le n$). We think of the curves $c_1, \ldots, c_n$ as being "linearly independent directions in the tangent space at m" (see Proposition 9.4.14). We call these curves the *standard basis curves at m determined by* ϕ. It is important to keep in mind that these curves depend on the choice of the chart ϕ; the notation $c_1, \ldots, c_n$ fails to remind us of this.

Proposition 9.4.11. *Let ϕ be a chart at a point m in an n-manifold. Then the map*

$$C_\phi \colon \tilde{T}_m \to \mathbb{R}^n \colon \tilde{c} \mapsto D(\phi \circ c)(0)$$

is well-defined and bijective.

Notice that initially we had no way of "adding" curves b and c at a point m in a manifold or of "multiplying" them by scalars. Now, however, we can use the bijection C_ϕ to transfer the vector space structure from $\mathbb{R}^n$ to the tangent space $\widetilde{T}_m$. Thus we add equivalence classes $\widetilde{b}$ and $\widetilde{c}$ in the obvious fashion. The formula is

$$\widetilde{b} + \widetilde{c} = C_\phi{}^{-1}\big(C_\phi(\widetilde{b}) + C_\phi(\widetilde{c})\big). \tag{9.1}$$

Similarly, if b is a curve at m and α is a scalar, then

$$\alpha\widetilde{b} = C_\phi{}^{-1}\big(\alpha C_\phi(\widetilde{b})\big). \tag{9.2}$$

Corollary 9.4.12. *At every point m in an n-manifold the geometric tangent space $\widetilde{T}_m$ may be regarded as a vector space isomorphic to $\mathbb{R}^n$.*

As remarked previously, we have defined the vector space structure of the geometric tangent space at m in terms of the mapping C_ϕ, which in turn depends on the choice of a particular chart ϕ. From this it might appear that addition and scalar multiplication on the tangent space depend on ϕ. Happily this is not so.

Proposition 9.4.13. *Let m be a point in an n-manifold. The vector space structure of the geometric tangent space $\widetilde{T}_m$ is independent of the particular chart ϕ used to define it.*

Proposition 9.4.14. *Let ϕ be a chart containing a point m in an n-manifold. Let $\mathbf{c}_1, \ldots, \mathbf{c}_n$ be the standard basis curves determined by ϕ (see Example 9.4.10). Then $\{\widetilde{\mathbf{c}}_1, \ldots, \widetilde{\mathbf{c}}_n\}$ is a basis for the tangent space $\widetilde{T}_m$.*

Exercise 9.4.15. We know from the preceding proposition that if $\widetilde{c}$ belongs to $\widetilde{T}_m$, then there exist scalars $\alpha_1, \ldots, \alpha_n$ such that $\widetilde{c} = \sum_{k=1}^n \alpha_k \widetilde{\mathbf{c}}_k$. Find these scalars.

If $F\colon M \to N$ is a smooth mapping between manifolds and m is a point in M, we denote by $\widetilde{d}F_m$ the mapping that takes each geometric tangent vector $\widetilde{c}$ in $\widetilde{T}_m$ to the corresponding geometric tangent vector $(F \circ c)^\sim$ in $\widetilde{T}_{F(m)}$. That $\widetilde{d}F_m$ is well-defined is clear from Proposition 9.4.7. The point of Proposition 9.4.17 below is to make the notation for this particular map plausible.

Definition 9.4.16. If $F\colon M \to N$ is a smooth mapping between manifolds and $m \in M$, then the function

$$\widetilde{d}F_m\colon \widetilde{T}m \to \widetilde{T}_{F(m)}\colon \widetilde{c} \mapsto (F \circ c)^\sim$$

is the DIFFERENTIAL of F at m.

Proposition 9.4.17. *Let $F \colon M \to N$ be a smooth map from an n-manifold M to a p-manifold N. For every $m \in M$ and every pair of charts ϕ at m and ψ at $F(m)$ the following diagram commutes*

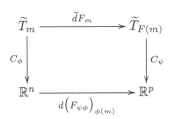

and consequently $\widetilde{d}F_m$ is a linear map. (The maps C_ϕ and C_ψ are defined as in Proposition 9.4.11.)

The map C_ϕ has been used to provide the geometric tangent space $\widetilde{T}_m$ at a point m on an n-manifold with the vector space structure of $\mathbb{R}^n$. With equal ease it may be used to provide $\widetilde{T}_m$ with a norm. By defining

$$\|\widetilde{c}\| = \|C_\phi(\widetilde{c})\|$$

it is clear that we have made $\widetilde{T}_m$ into a normed linear space (but in a way that *does* depend on the choice of the chart ϕ). Furthermore, under this definition C_ϕ is an isometric isomorphism between $\widetilde{T}_m$ and $\mathbb{R}^n$. Thus, in particular, we may regard T_m and $T_{F(m)}$ as Euclidean spaces. From the preceding proposition we see that the map $\widetilde{d}F_m$ is (continuous and) linear, being the composite of the mapping $d\big(F_{\phi\psi}\big)_{\phi(m)}$ with two isometric isomorphisms $\big(C_\psi\big)^{-1}$ and C_ϕ.

If we use the mapping C_ϕ to identify $\widetilde{T}_m$ and $\mathbb{R}^n$ as Banach spaces and, similarly, C_ψ to identify $\widetilde{T}_{F(m)}$ and $\mathbb{R}^p$, then the continuous linear maps $\widetilde{d}F_m$ and $d\big(F_{\phi\psi}\big)_{\phi(m)}$ are also identified. The notation that we have used for the mapping $\widetilde{d}F_m$ is thus a reasonable one since, as we have just seen, this mapping can be identified with the differential at $\phi(m)$ of the local representative of F and since its *definition* does not depend on the charts ϕ and ψ. To further strengthen the case for the plausibility of this notation. Consider what happens if M and N are open subsets of $\mathbb{R}^n$ and $\mathbb{R}^p$, respectively (regarded as manifolds whose differential structure is generated in each case by the appropriate identity map). The corresponding local representative of F is F itself, in which case the bottom map in the diagram for Proposition 9.4.17 is simply dF_m.

The preceding definition also helps to justify the usual intuitive picture for familiar n-manifolds of the tangent space being a copy of $\mathbb{R}^n$ placed at m. Suppose that for a particular n-manifold M there exists a smooth inclusion map of M into a higher dimensional Euclidean space $\mathbb{R}^p$. (For example, the inclusion map of the 2-manifold $\mathbb{S}^2$ into $\mathbb{R}^3$ is smooth.) Then taking F in the preceding discussion to be this inclusion map, the function $\widetilde{d}F_m$ maps $\widetilde{T}_m$ into a subspace of $\widetilde{T}_{F(m)} = \mathbb{R}^p$. Picture this subspace being translated to the point $m = F(m)$ in $\mathbb{R}^p$. To see how this works it is best to work through the details of some examples.

Example 9.4.18. Let $U = \{(x, y, z) \in \mathbb{S}^2 : x > 0\}$. The map $\phi : U \to \mathbb{R}^2 : (x, y, z) \mapsto (y, z)$ is a chart for the 2-manifold $\mathbb{S}^2$. Let $\iota : \mathbb{S}^2 \to \mathbb{R}^3$ be the inclusion map of $\mathbb{S}^2$ into $\mathbb{R}^3$ and let $m = \left(\frac{1}{2}, \frac{1}{\sqrt{2}}, \frac{1}{2} \right) \in \mathbb{S}^2$. Then the range of $d(\iota_\phi)_{\phi(m)}$ is the plane in $\mathbb{R}^3$ whose equation is $x + \sqrt{2}y + z = 0$. If we translate this plane to the point m, we obtain the plane whose equation is $x + \sqrt{2}y + z = 2$, which is exactly the result we obtain using techniques of beginning calculus to "find the equation of the tangent plane to the surface $x^2 + y^2 + z^2 = 1$ at the point $\left(\frac{1}{2}, \frac{1}{\sqrt{2}}, \frac{1}{2} \right)$".

Example 9.4.19. Same example as the preceding except this time let $U = \{(x, y, z) \in \mathbb{S}^2 : z \neq -1\}$ and $\phi : U \to \mathbb{R}^2$ be the stereographic projection of $\mathbb{S}^2$ from the south pole. That is,

$$\phi(x, y, z) = \left(\frac{x}{1 + z}, \frac{y}{1 + z} \right).$$

Proposition 9.4.20 (A Chain Rule for Maps Between Manifolds). *If $F : M \to N$ and $G : N \to P$ are smooth mappings between manifolds, then*

$$\widetilde{d}(G \circ F)_m = \widetilde{d}G_{F(m)} \circ \widetilde{d}F_m \tag{9.3}$$

for every $m \in M$.

9.5. The Algebraic Tangent Space

There is another way of looking at the tangent space at a point m in a manifold. Instead of regarding tangent vectors as "directions" determined by equivalence classes of curves at m, we will now consider them to be "directional derivatives" of (germs of) smooth functions defined in neighborhoods of m.

Definition 9.5.1. Let m be a point in a manifold M and $f, g \in \mathcal{C}_m^\infty(M, \mathbb{R})$ (the family of all smooth real valued functions defined on a neighborhood of m). We write $f \sim g$ if there exists a neighborhood of m on which f and g agree. Then $\sim$ is clearly an equivalence relation on $\mathcal{C}_m^\infty(M, \mathbb{R})$. The corresponding equivalence classes are GERMS of smooth functions at m. If f is a member of $\mathcal{C}_m^\infty(M, \mathbb{R})$, we denote the germ containing f by $\widehat{f}$. The family of all germs of smooth real valued functions at m is denoted by $\mathcal{G}_m(M)$ (or just $\mathcal{G}_m$). Addition, multiplication, and scalar multiplication of germs are defined as you would expect. For $\widehat{f}, \widehat{g} \in \mathcal{G}_m$ and $\alpha \in \mathbb{R}$ let

$$\widehat{f} + \widehat{g} = (f + g)^{\widehat{}}$$
$$\widehat{f}\,\widehat{g} = (fg)^{\widehat{}}$$
$$\alpha\widehat{f} = (\alpha f)^{\widehat{}}$$

(As usual, the domain of $f + g$ and fg is taken to be $\operatorname{dom} f \cap \operatorname{dom} g$.)

Proposition 9.5.2. *If m is a point in a manifold, then the set $\mathcal{G}_m$ of germs of smooth functions at m is (under the operations defined above) a unital commutative algebra.*

Definition 9.5.3. Let m be a point in a manifold. A DERIVATION on the algebra $\mathcal{G}_m$ of germs of smooth functions at m is a linear functional $v \in \mathcal{G}_m^*$ that satisfies *Leibniz's rule*

$$v(\widehat{f}\,\widehat{g}) = f(m)v(\widehat{g}) + v(\widehat{f})g(m)$$

for all $\widehat{f}, \widehat{g} \in \mathcal{G}_m$. Another name for a derivation on the algebra $\mathcal{G}_m$ is an ALGEBRAIC TANGENT VECTOR at m. The set of all algebraic tangent vectors at m (that is, derivations on $\mathcal{G}_m$) will be called the ALGEBRAIC TANGENT SPACE at m and will be denoted by $\widehat{T}_m(M)$ (or just $\widehat{T}_m$).

The idea here is to bring to manifolds the concept of *directional derivative*; as the next example shows directional derivatives at points in a normed linear space are derivations on the algebra of (germs of) smooth functions.

Example 9.5.4. Let V be a normed linear space and let a and v be vectors in V. Define $D_{v,a}$ on functions f in $\mathcal{G}_a(V)$ by

$$D_{v,a}(\widehat{f}) := D_v f(a).$$

(Here, $D_v f(a)$ is the usual directional derivative of f at a in the direction of v from beginning calculus.) Then $D_{v,a}$ is well-defined and is a derivation on $\mathcal{G}_a$. (Think of the operator $D_{v,a}$ as being "differentiation in the direction of the vector v followed by evaluation at a.")

Proposition 9.5.5. *If m is a point in a manifold, v is a derivation on $\mathcal{G}_m$, and k is a smooth real valued function constant in some neighborhood of m, then $v(\widehat{k}) = 0$.*

Hint for proof. Recall that $1 \cdot 1 = 1$.

It is an easy matter to show that the terminology "algebraic tangent space" adopted in Definition 9.5.3 is not overly optimistic. It *is* a vector space.

Proposition 9.5.6. *If m is a point in a manifold, then the tangent space $\widehat{T}_m$ is a vector space under the usual pointwise definition of addition and scalar multiplication.*

We now look at an example that establishes a first connection between the geometric and the algebraic tangent spaces.

Example 9.5.7. Let m be a point in a manifold M and $\widetilde{c}$ be a vector in the geometric tangent space $\widetilde{T}_m$. Define

$$v_{\widetilde{c}} \colon \mathcal{G}_m \to \mathbb{R} \colon \widehat{f} \mapsto D(f \circ c)(0).$$

Then $v_{\widetilde{c}}$ is well-defined and belongs to the algebraic tangent space $\widehat{T}_m$.

The notation $v_{\widetilde{c}}(\widehat{f})$ of the preceding example is not particularly attractive. In the following material we will ordinarily write just $v_c(f)$. Although strictly speaking this is incorrect, it should not lead to confusion. We have shown that $v_{\widetilde{c}}(\widehat{f})$ depends *only* on the equivalence classes $\widetilde{c}$ and $\widehat{f}$, not on the representatives chosen. Thus we do not distinguish between $v_b(g)$ and $v_c(f)$ provided that b and c belong to the same member of $\widetilde{T}_m$ and f and g to the same germ.

Example 9.5.7 is considerably more general than it may at first appear. Later in this section we will show that the association $\widetilde{c} \mapsto v_{\widetilde{c}}$ is an isomorphism between the tangent spaces $\widetilde{T}_m$ and $\widehat{T}_m$. In particular, there are no derivations on $\mathcal{G}_m$ other than those induced by curves at m.

For the moment, however, we wish to make plausible for an n-manifold the use of the notation $\frac{\partial}{\partial x^k}\big|_m$ for the derivation $v_{\mathbf{c}_k}$, where $\mathbf{c}_k$ is the kth standard basis curve at the point m determined by the chart $\phi = (x^1, \ldots, x^n)$ (see the notational convention in 9.2.5 and Example 9.4.10). The crux of the matter is the following proposition, which says that if $c_{\mathbf{u}} = \phi^{-1} \circ b_{\mathbf{u}}$, where ϕ is a chart at m and $b_{\mathbf{u}}$ is the parametrized line through $\phi(m)$ in the direction of the vector $\mathbf{u}$, then the value at a germ $\widehat{f}$ of the derivation

v_{c_u} may be found by taking the directional derivative in the direction $\mathbf{u}$ of the local representative f_ϕ.

Proposition 9.5.8. *Let ϕ be a chart containing the point m in an n-manifold. If $\mathbf{u} \in \mathbb{R}^n$ and $b_\mathbf{u} : \mathbb{R} \to \mathbb{R}^n$:*
$t \mapsto \phi(m) + t\mathbf{u}$, *then $c_\mathbf{u} := \phi^{-1} \circ b_\mathbf{u}$ is a curve at m and*

$$v_{c_\mathbf{u}}(f) = D_\mathbf{u}(f_\phi)(\phi(m))$$

for all $\widehat{f} \in \mathcal{G}(m)$.

Hint for proof. Use Proposition 25.5.9 in [12].

Definition 9.5.9. In Proposition 9.4.14 we saw that if ϕ is a chart containing a point m in an n-manifold, then the vectors $\widetilde{\mathbf{c}}_1, \ldots, \widetilde{\mathbf{c}}_n$ form a basis for the geometric tangent space $\widetilde{T}_m$. We call these vectors the BASIS VECTORS OF $\widetilde{T}_m$ DETERMINED BY ϕ.

Corollary 9.5.10. *Let ϕ be a chart containing the point m in an n-manifold. If $\widetilde{\mathbf{c}}_1, \ldots, \widetilde{\mathbf{c}}_n$ are the basis vectors for $\widetilde{T}_m$ determined by ϕ, then*

$$v_{\mathbf{c}_k}(f) = \left(f_\phi \right)_k (\phi(m)) \tag{9.4}$$

for every $\widehat{f}$ in $\mathcal{G}_m$. (The subscript k on the right side of the equation indicates differentiation; $\left(f_\phi \right)_k$ is the kth partial derivative of the local representative f_ϕ, in another notation the right hand side is $\frac{\partial f_\phi}{\partial x^k}(\phi(m))$.)

Notation 9.5.11. The preceding corollary says that (in an n-manifold M) the action of the derivation $v_{\mathbf{c}_k}$ on a function f (technically, on the germ $\widehat{f}$) is that of partial differentiation of the local representative f_ϕ followed by evaluation at the point $\phi(m)$. In particular, if M happens to be an open subset of $\mathbb{R}^n$, then (9.4) becomes

$$v_{\mathbf{c}_k}(f) = f_k(m)$$

so that the value of $v_{\mathbf{c}_k}$ at f *is* the kth partial derivative of f (evaluated at m). It is helpful for the notation to remind us of this fact; so in the following material we will usually write $\left. \frac{\partial}{\partial x^k} \right|_m$ for $v_{\mathbf{c}_k}$, where $1 \le k \le n$ and $\phi = (x^1, \ldots, x^n)$ is the chart containing m in terms of which the curves $\mathbf{c}_k$ are defined. The value of this derivation at f will be denoted by $\frac{\partial f}{\partial x^k}(m)$. Thus

$$\frac{\partial f}{\partial x^k}(m) = \left. \frac{\partial}{\partial x^k} \right|_m (f) = v_{\mathbf{c}_k}(f) = v_{\widetilde{\mathbf{c}}_k}(\widehat{f}).$$

Let $\phi = (x_1, \ldots, x^n)$ and $\psi = (y^1, \ldots, y^n)$ be charts containing a point m on an n-manifold. It is perhaps tempting, but utterly wrong, to believe that if $x^k = y^k$ for some k, then $\frac{\partial}{\partial x^k}\big|_m = \frac{\partial}{\partial y^k}\big|_m$. The formula

$$\frac{\partial f}{\partial x^k}(m) = (f \circ \phi^{-1})_k(\phi(m))$$

(see 9.5.10 and 9.5.11) should make it clear that $\frac{\partial}{\partial x^k}$ depends on *all* the components of the chart ϕ and not just on the single component x^k. In any case, here is a concrete counterexample.

Example 9.5.12. Consider $\mathbb{R}^2$ as a 2-manifold (with the usual differential structure generated by the atlas whose only member is the identity map on $\mathbb{R}^2$). Let $\phi = (x^1, x^2)$ be the identity map on $\mathbb{R}^2$ and $\psi = (y^1, y^2)$ be the map defined by $\psi \colon \mathbb{R}^2 \to \mathbb{R}^2 \colon (u, v) \mapsto (u, u + v)$. Clearly ϕ and ψ are charts containing the point $m = (1, 1)$ and $x^1 = y^1$. We see, however, that $\frac{\partial}{\partial x^1} \neq \frac{\partial}{\partial y^1}$ by computing $\frac{\partial f}{\partial x^1}(1, 1)$ and $\frac{\partial f}{\partial y^1}(1, 1)$ for the function $f \colon \mathbb{R}^2 \to \mathbb{R} \colon (u, v) \mapsto u^2 v$.

Proposition 9.5.13 (Change of Variables Formula). *Let* $\phi = (x^1, \ldots, x^n)$ *and* $\psi = (y^1, \ldots, y^n)$ *be charts at a point* m *on an* n-*manifold. Then*

$$\frac{\partial}{\partial x^k}\bigg|_m = \sum_{j=1}^n \frac{\partial y^j}{\partial x^k}(m) \frac{\partial}{\partial y^j}\bigg|_m \qquad (9.5)$$

for $1 \leq k \leq n$.

Remark 9.5.14. If $\phi = (x^1, \ldots, x^n)$ and $\psi = (y^1, \ldots, y^n)$ are charts on an n-manifold which overlap (that is, $\operatorname{dom}\phi \cap \operatorname{dom}\psi \neq \emptyset$), then the preceding *change of variables formula* holds for *all* m in $\operatorname{dom}\phi \cap \operatorname{dom}\psi \neq \emptyset$. Thus in the interests of economy of notation, symbols indicating evaluation at m are normally omitted. Then (9.5) becomes

$$\frac{\partial}{\partial x^k} = \sum_{j=1}^n \frac{\partial y^j}{\partial x^k} \frac{\partial}{\partial y^j} \qquad (9.6)$$

with the understanding that it may be applied to any real valued function which is smooth on $\operatorname{dom}\phi \cap \operatorname{dom}\psi$.

Exercise 9.5.15. Regard the 2-sphere in $\mathbb{R}^3$ as a 2-manifold whose differentiable structure is is generated by the stereographic projections from the north and south poles (see Example 9.2.12). Let ϕ be the stereographic projection from the south pole. That is,

$$\phi(x, y, z) = \left(\frac{x}{1+z}, \frac{y}{1+z} \right)$$

for all $(x, y, z) \in \mathbb{S}^2$ such that $z \neq -1$. Let ψ be defined by

$$\psi(x, y, z) = (y, z)$$

for all $(x, y, z) \in \mathbb{S}^2$ such that $x > 0$. (It is easy to see that ψ is a chart.) Also define a real valued function f by

$$f(x, y, z) = \frac{x}{y} + \frac{y}{z}$$

for all $(x, y, z) \in \mathbb{S}^2$ such that x, y, $z > 0$. Let $m = \left(\frac{1}{2}, \frac{1}{\sqrt{2}}, \frac{1}{2} \right)$.

For these data verify by explicit computation Equation (9.5). That is, by computing both sides show that for $k = 1$ and $k = 2$ the formula

$$\frac{\partial f}{\partial x^k}(m) = \frac{\partial y^1}{\partial x^k}(m) \frac{\partial f}{\partial y^1}(m) + \frac{\partial y^2}{\partial x^k}(m) \frac{\partial f}{\partial y^2}(m)$$

is correct for the functions ϕ, ψ, and f and the point m given in the preceding paragraph.

Proposition 9.5.16 (Another version of the chain rule). *Let* $G\colon M \to N$ *be a smooth map between manifolds of dimensions n and p, respectively, $\phi = (x^1, \ldots, x^n)$ be a chart containing the point m in M, and $\psi = (y^1, \ldots, y^p)$ be a chart containing $G(m)$ in N. Then*

$$\frac{\partial (f \circ G)}{\partial x^k}(m) = \sum_{j=1}^{p} \frac{\partial (y^j \circ G)}{\partial x^k}(m) \frac{\partial f}{\partial y^j}(G(m)) \qquad (9.7)$$

whenever $f \in \mathcal{C}^\infty_{G(m)}$ and $1 \leq k \leq n$.

Remark 9.5.17. If one is willing to adopt a sufficiently relaxed attitude towards notation many complicated looking formulas can be put in "simpler" form. Convince yourself that it is not beyond the realm of possibility for one to encounter equation (9.7) written in the form

$$\frac{\partial z}{\partial x^k} = \sum_{j=1}^{p} \frac{\partial z}{\partial y^j} \frac{\partial y^j}{\partial x^k}.$$

Exercise 9.5.18. Consider the 2-manifold $\mathbb{S}^2$ with the differential structure generated by the stereographic projections form the north and south poles (see Example 9.2.12) and the 2-manifold $\mathbb{P}^2$ with the differentiable structure generated by the atlas given in Example 9.2.14. Recall from Example 9.3.5 that the map $F\colon \mathbb{S}^2 \to \mathbb{P}^2\colon (x, y, z) \mapsto [(x, y, z)]$ is smooth. Define

$$h\big([(x, y, z)]\big) = \frac{x + y}{2z}$$

whenever $[(x, y, z)] \in \mathbb{P}^2$ and $z \neq 0$; and let $m = \big(\frac{1}{2}, \frac{1}{\sqrt{2}}, \frac{1}{2}\big)$. It is clear that h is well-defined. (You may assume it is smooth in a neighborhood of $F(m)$.) Let $\phi = (u, v)$ be the stereographic projection of $\mathbb{S}^2$ from the south pole (see 9.2.12).

(a) Using the *definitions* of $\frac{\partial}{\partial u}\big|_m$ and $\frac{\partial}{\partial v}\big|_m$ compute $\frac{\partial(h \circ F)}{\partial u}(m)$ and $\frac{\partial(h \circ F)}{\partial v}(m)$.

(b) Let η be the chart in $\mathbb{P}^2$ defined in Example 9.2.14. Use this chart, which contains $F(m)$, and the version of the *chain rule* given in Proposition 9.5.16 to compute (independently of part (a)) $\frac{\partial(h \circ F)}{\partial u}(m)$ and $\frac{\partial(h \circ F)}{\partial v}(m)$.

Let m be a point in an n-manifold. In Example 9.5.7 we defined, for each $\tilde{c}$ in the geometric tangent space $\tilde{T}_m$, a function

$$v_{\tilde{c}}\colon \mathcal{G}_m \to \mathbb{R}\colon \hat{f} \mapsto D(f \circ c)(0)$$

and showed that $v_{\tilde{c}}$ is (well-defined and) a derivation on the space $\mathcal{G}_m$ of germs at m. Thus the map $v\colon \tilde{c} \mapsto v_{\tilde{c}}$ takes members of $\tilde{T}_m$ to members of $\hat{T}_m$. The next few propositions lead to the conclusion that $v\colon \tilde{T}_m \to \hat{T}_m$ is a vector space isomorphism and that consequently the two definitions of "tangent space" are essentially the same. Subsequently we will drop the diacritical marks tilde and circumflex that we have used to distinguish the geometric and algebraic tangent spaces and instead write just T_m for the tangent space at m. We will allow context to dictate whether a tangent vector (that is, a member of the tangent space) is to be interpreted as an equivalence class of curves or as a derivation. Our first step is to show that the map v is linear.

Proposition 9.5.19. *Let m be a point in an n-manifold and*

$$v\colon \tilde{T}_m \to \hat{T}_m\colon \tilde{c} \mapsto v_{\tilde{c}}$$

be the map defined in Example 9.5.7. Then

(i) $v_{\widetilde{b}+\widetilde{c}} = v_{\widetilde{b}} + v_{\widetilde{c}}$ and

(ii) $v_{\alpha\widetilde{c}} = \alpha v_{\widetilde{c}}$

for all $\widetilde{b},\ \widetilde{c} \in \widetilde{T}_m$ and $\alpha \in \mathbb{R}$.

Proposition 9.5.20. *Let m be a point in an n-manifold. Then the map*

$$v\colon \widetilde{T}_m \to \widehat{T}_m \colon \widetilde{c} \mapsto v_{\widetilde{c}}$$

(defined in 9.5.7) is injective.

In order to show that the map $v\colon \widetilde{T}_m \to \widehat{T}_m$ is surjective we will need to know that the tangent vectors $\frac{\partial}{\partial x^1}\big|_m, \ldots, \frac{\partial}{\partial x^n}\big|_m$ span the tangent space $\widehat{T}_m$. The crucial step in this argument depends on adapting the (second order) *Taylor's formula* so that it holds on finite dimensional manifolds.

Lemma 9.5.21. *Let $\phi = (x^1, \ldots, x^n)$ be a chart containing a point m in an n-manifold and f be a member of C_m^∞. Then there exist a neighborhood U of m and smooth functions s^{jk} (for $1 \le j, k \le n$) such that*

$$f = f(m) + \sum_{j=1}^{n}(x^j - a^j)\big(f_\phi\big)_j(a) + \sum_{j,k=1}^{n}(x^j - a^j)(x^k - a^k)s^{jk}$$

where $a = \phi(m)$.

Hint for proof. Apply *Taylor's formula* to the local representative f_ϕ.

Proposition 9.5.22. *If $\phi = (x^1, \ldots, x^n)$ is a chart containing a point m in an n-manifold, then the derivations $\frac{\partial}{\partial x^1}\big|_m, \ldots, \frac{\partial}{\partial x^n}\big|_m$ span the algebraic tangent space $\widehat{T}_m$. In fact, if w is an arbitrary element of $\widehat{T}_m$, then*

$$w = \sum_{j=1}^{n} w\big(\widehat{x^j}\big)\frac{\partial}{\partial x^j}\bigg|_m.$$

Proposition 9.5.23. *Let m be a point in an n-manifold. Then the map*

$$v\colon \widetilde{T}_m \to \widehat{T}_m \colon \widetilde{c} \mapsto v_{\widetilde{c}}$$

(defined in 9.5.7) is surjective.

Corollary 9.5.24. *Let m be a point in an n-manifold. Then the map v (of the preceding proposition) is an isomorphism between the tangent spaces $\widetilde{T}_m$ and $\widehat{T}_m$.*

Corollary 9.5.25. *If $\phi = (x^1, \ldots, x^n)$ is a chart containing a point m in an n-manifold, then the derivations $\frac{\partial}{\partial x^1}\big|_m, \ldots, \frac{\partial}{\partial x^n}\big|_m$ constitute a basis for the tangent space $\widehat{T}_m$.*

In 9.4.16 we defined the differential $\widetilde{d}F_m$ of a smooth map $F\colon M \to N$ between finite dimensional manifolds at a point $n \in M$. This differential between the geometric tangent spaces at m and $F(m)$ turned out to be a linear map (see 9.4.17). In a similar fashion F induces a linear map, which we denote by $\widehat{d}F_m$, between the algebraic tangent spaces at m and $F(m)$. We define this new *differential* and then show that it is essentially same as the one between the corresponding geometric tangent spaces.

Definition 9.5.26. Let $F\colon M \to N$ be a smooth map between finite dimensional manifolds, $m \in M$, and $w \in \widehat{T}_m$. Define $\widehat{d}F_m\colon \widehat{T}_m \to \widehat{T}_{F(m)}$ by setting $\big(\widehat{d}F_m(w)\big)(\widehat{g}) = w(\widehat{g \circ F})$ — or in somewhat less cluttered notation

$$\widehat{d}F_m(w)(g) = w(g \circ F)$$

for each $g \in \mathcal{C}^\infty_{F(m)}$.

Proposition 9.5.27. *The function $\widehat{d}F_m(w)$, defined above, is well-defined and is a derivation on $\mathcal{G}_{F(m)}$.*

Now we show that this new differential is essentially the same as the one defined in 9.4.16.

Proposition 9.5.28. *Let $F\colon M \to N$ be a smooth mapping between finite dimensional manifolds and $m \in M$. Then the following diagram commutes.*

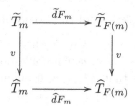

In light of the isomorphism between the geometric and algebraic tangent spaces to a manifold (see 9.5.24) and the equivalence of the respective differential maps (proved in the preceding proposition), we will for the most part write just T_m for either type of tangent space and dF_m for either differential of a smooth map. In situations where the difference is important context should make it clear which one is intended.

Corollary 9.5.29. *If $F\colon M \to N$ is a smooth map between finite dimensional manifolds and $m \in M$, then $\widehat{d}F_m$ is a linear transformation from the algebraic tangent space $\widehat{T}_m$ into $\widehat{T}_{F(m)}$.*

Proposition 9.5.30 (Yet another chain rule). *If $F\colon M \to N$ and $G\colon N \to P$ are smooth maps between finite dimensional manifolds and $m \in M$, then*

$$\widehat{d}(G \circ F)_m = \widehat{d}G_{F(m)} \circ \widehat{d}F_m.$$

In the next exercise we consider the tangent space T_a at a point a in $\mathbb{R}$. Here, as usual, the differential structure on $\mathbb{R}$ is taken to be the one generated by the atlas whose only chart is the identity map I on $\mathbb{R}$. The tangent space is one-dimensional; it is generated by the tangent vector $\frac{\partial}{\partial I}\big|_a$. This particular notation is, as far as I know, never used; some alternative standard notations are $\frac{d}{dx}\big|_a$, $\frac{d}{dt}\big|_a$, and $\frac{d}{dI}\big|_a$. And, of course, $\frac{d}{dx}\big|_a (f)$ is written as $\frac{df}{dx}(a)$.

Exercise 9.5.31. Let $a \in \mathbb{R}$ and $g \in \mathcal{C}_a^\infty$. Find $\frac{dg}{dx}(a)$.

Exercise 9.5.32. If $f \in \mathcal{C}_a^\infty$, where m is a point in some n-manifold, and $w \in T_m$, then $df_m(w)$ belongs to $T_{f(m)}$. Since the tangent space at $f(m)$ is one-dimensional, there exists $\lambda \in \mathbb{R}$ such that $df_m(w) = \lambda \frac{d}{dI}\big|_{f(m)}$. Show that $\lambda = w(f)$.

Convention 9.5.33. Let a be a point in $\mathbb{R}$. It is natural in the interests of simplifying notation to make use of the isomorphism $\lambda \frac{d}{dI}\big|_a \mapsto \lambda$ between the tangent space T_a and $\mathbb{R}$ to identify these one-dimensional spaces. If $f \in \mathcal{C}_m^\infty$ where m is a point in an n-manifold and if we regard $T_{f(m)} = \mathbb{R}$, then Corollary 9.5.29 says that df_m is a linear map from T_m into $\mathbb{R}$; that is, df_m belongs to the dual space of T_m; that is

$$df_m \in T_m{}^*. \tag{9.8}$$

Furthermore, under the identification $\lambda = \lambda \frac{d}{dI}\big|_{f(m)}$ we conclude from Exercise 9.5.32 that

$$df_m(w) = w(f). \tag{9.9}$$

for every $w \in T_m$. From now on we adopt (9.8) and (9.9) even though they are, strictly speaking, abuses of notation. They should cause little confusion and are of considerable help in reducing notational clutter.

Definition 9.5.34. If m is a point in an n-manifold, the dual space $T_m{}^*$ of the tangent space at m is called the COTANGENT SPACE at m.

Notice that in (9.8) we have adopted the convention that at every point m the differential of a smooth real valued function f belongs to the cotangent space at m. In particular, if $\phi = (x^1, \ldots, x^n)$ is a chart on an n-manifold, then each of its components x^k is a smooth real valued function and therefore $dx^k{}_m$ belongs to the cotangent space $T_m{}^*$ for every m in the domain of ϕ. The next proposition shows that in fact the set $\{dx^1{}_m, \ldots, dx^n{}_m\}$ of cotangent vectors is the basis for $T_m{}^*$ dual to the basis for T_m given in Corollary 9.5.25.

Proposition 9.5.35. Let $\phi = (x^1, \ldots, x^n)$ be a chart containing the point m in an n-manifold. Then $\{dx^1{}_m, \ldots, dx^n{}_m\}$ is a basis for $T_m{}^*$; it is dual to the basis $\left\{\frac{\partial}{\partial x^1}\big|_m, \ldots, \frac{\partial}{\partial x^n}\big|_m\right\}$ for T_m.

Proposition 9.5.36. Let m be a point in an n-manifold, $f \in \mathcal{C}_m^\infty$, and $\phi = (x^1, \ldots, x^n)$ be a chart containing m. Then

$$df_m = \sum_{k=1}^n \frac{\partial f}{\partial x^k}(m)\, dx^k{}_m.$$

Hint for proof. There exist scalars $\alpha_1, \ldots, \alpha_n$ such that $df_m = \sum_{j=1}^n \alpha_j\, dx^j{}_m$. (Why?) Consider $\sum_{j=1}^n \alpha_j\, dx^j{}_m\left(\frac{\partial}{\partial x^k}\big|_m\right)$.

Notice that this proposition provides some meaning (and justification) for the conventional formula frequently trotted out in beginning calculus courses.

$$df = \frac{\partial f}{\partial x}\, dx + \frac{\partial f}{\partial y}\, dy + \frac{\partial f}{\partial z}\, dz.$$

Chapter 10

DIFFERENTIAL FORMS ON MANIFOLDS

In everything that follows all vector spaces are assumed to be real, finite dimensional, and oriented; and all manifolds are smooth oriented differentiable manifolds.

10.1. Vector Fields

Definition 10.1.1. Let M be a manifold. The set

$$TM := \bigcup_{m \in M} T_m$$

is the TANGENT BUNDLE of M. The TANGENT BUNDLE PROJECTION is the map $\tau_M \colon TM \to M$ such that $\tau_M(w) = m$ whenever $w \in T_m$.

Definition 10.1.2. A VECTOR FIELD on a manifold M is a map $v \colon M \to TM$ such that $\tau_M \circ v = Id_M$ (the identity map on M). It is an equivalent formulation to say that v is a vector field if it is a right inverse of the tangent bundle projection or to say that it is a *section* of the tangent bundle.

Notation 10.1.3. For a manifold M denote by $\mathcal{C}^\infty(M)$ the family of smooth real valued functions on M. That is, a function f belongs to $\mathcal{C}^\infty(M)$ provided that it belongs to $\mathcal{C}^\infty_m(M, \mathbb{R})$ for every $m \in M$ (see Definition 9.5.1).

Definition 10.1.4. If v is a vector field on a manifold M and $f \in \mathcal{C}^\infty(M)$, define the function $vf \colon M \to \mathbb{R}$ by

$$(vf)(m) = v(m)(\hat{f}).$$

We will say that v is a SMOOTH vector field if vf is smooth for every $f \in \mathcal{C}^\infty(M)$.

Exercise 10.1.5. Let M be an n-manifold and $\phi = (x^1, \ldots, x^n) \colon U \to \mathbb{R}^n$ be a chart. Regard U as a submanifold of M in the obvious fashion. For $1 \le k \le n$ take $\frac{\partial}{\partial x^k}$ to be the vector field on U defined by

$$\frac{\partial}{\partial x^k} \colon m \mapsto \left. \frac{\partial}{\partial x^k} \right|_m$$

where, as before (see 9.5.10 and 9.5.11),

$$\left. \frac{\partial}{\partial x^k} \right|_m (\hat{f}) = \left(f \circ \phi^{-1} \right)_k (\phi(m))$$

for all $f \in \mathcal{C}^\infty(U)$. Show that the vector field $\frac{\partial}{\partial x^k}$ is smooth.

Exercise 10.1.6. Notation as in the preceding exercise. Let v be a smooth vector field on U. For every $m \in U$ there exist numbers $\alpha_1(m), \ldots, \alpha_n(m)$ such that

$$v(m) = \sum_{k=1}^{n} \alpha_k(m) \left. \frac{\partial}{\partial x^k} \right|_m$$

(see 9.5.25). Thus we may write

$$v = \sum_{k=1}^{n} \alpha_k \frac{\partial}{\partial x^k}$$

where α_k is the function $m \mapsto \alpha_k(m)$. Show that each α_k is a smooth function.

10.2. Differential 1-forms

Definition 10.2.1. Let M be a manifold. The set

$$T^*M := \bigcup_{m \in M} T_m{}^*$$

is the COTANGENT BUNDLE of M. The COTANGENT BUNDLE PROJECTION is the map $\rho_M \colon T^*M \to M$ such that $\rho_M(u) = m$ whenever $u \in T_m{}^*$.

Definition 10.2.2. A DIFFERENTIAL ONE-FORM (or DIFFERENTIAL 1-FORM) on a manifold M is a map $\omega\colon M \to T^*M$ such that $\rho_M \circ \omega = Id_M$ (the identity map on M). Equivalently, it is a right inverse of the cotangent bundle projection, or it is a section of the cotangent bundle.

Definition 10.2.3. If ω is a differential one-form on an n-manifold M and v is a vector field on M, we define

$$\omega(v)\colon M \to \mathbb{R}\colon m \mapsto \big(\omega(m)\big)\big(v(m)\big).$$

Definition 10.2.4. A differential one-form ω is SMOOTH if $\omega(v) \in \mathcal{C}^\infty(M)$ whenever v is a smooth vector field on M.

Proposition 10.2.5. *Let M be an n-manifold and $f \in \mathcal{C}^\infty(M)$. Then the function $df\colon m \mapsto df_m$ is a smooth differential one-form on M.*

Convention 10.2.6. From now on let's drop the words "smooth" and "differential" in the phrase *smooth differential one-form*. There is no other kind of "one-form" that we will be interested in.

10.3. Differential k-forms

Notation 10.3.1. For an n-manifold M and $k \geq 0$ let

$$\textstyle\bigwedge^k(M) = \bigcup\{\bigwedge^k(T_m{}^*)\colon m \in M\}$$

and

$$\textstyle\bigwedge(M) = \bigcup\{\bigwedge(T_m{}^*)\colon m \in M\}.$$

Definition 10.3.2. A DIFFERENTIAL FORM is a section of $\bigwedge(M)$. Thus ω is a differential form if $\omega(m) \in \bigwedge(T_m{}^*)$ for every $m \in M$. Similarly, ω is a DIFFERENTIAL k-FORM (or just a k-FORM) if it is a section of $\bigwedge^k(M)$. Notice that for 1-forms this definition agrees with the one given in 10.2.2 since $\bigwedge^1(M) = T^*M$. Also notice that a 0-form is just a real valued function on M (because of our identification of $\bigwedge^0(T_m{}^*)$ with $\mathbb{R}$ — see 8.1.15).

Exercise 10.3.3. Let M be a 3-manifold and $\phi = (x, y, z)\colon U \to \mathbb{R}^3$ be a chart. Regard U as a submanifold of M. Exhibit bases for $\bigwedge^0(U)$, $\bigwedge^1(U)$, $\bigwedge^2(U)$, $\bigwedge^3(U)$, and $\bigwedge(U)$.

Definition 10.3.4. Given a chart $\phi = (x^1, \ldots, x^n) \colon U \to \mathbb{R}^n$ on an n-manifold M, we may express a k-form ω locally (that is, on U) by

$$\omega(m) = \sum_{j_1 < \cdots < j_k} a_{j_1 \ldots j_k}(m) \, dx^{j_1}{}_m \wedge \cdots \wedge dx^{j_k}{}_m$$

for all $m \in U$. More succinctly,

$$\omega = \sum_{j_1 < \cdots < j_k} a_{j_1 \ldots j_k} \, dx^{j_1} \wedge \cdots \wedge dx^{j_k}.$$

It should be kept in mind that the "coefficients" $a_{j_1 \ldots j_k}$ in this expression are *functions* and that they depend on the choice of coordinate system (chart). The k-form ω is SMOOTH with respect to the chart ϕ if all the coefficients $a_{j_1 \ldots j_k}$ are smooth real valued functions on U. A k-form ω defined on all of M is SMOOTH if it is smooth with respect to every chart on M. A differential form is SMOOTH if its component in $\bigwedge^k(M)$ is smooth for every k. The set of smooth differential forms on M is denoted by $\mathcal{C}^\infty(M, \bigwedge(M))$ and the set of smooth k-forms by $\mathcal{C}^\infty(M, \bigwedge^k(M))$.

Convention 10.3.5. In the sequel all k-forms are smooth differential k-forms and all differential forms are smooth.

The next theorem defines a mapping d on differential forms called the EXTERIOR DIFFERENTIATION OPERATOR.

Theorem 10.3.6. *If M is an n-manifold, then there exists a unique linear map*

$$d \colon \mathcal{C}^\infty(M, \textstyle\bigwedge(M)) \to \mathcal{C}^\infty(M, \textstyle\bigwedge(M))$$

that satisfies

(1) $d^{\to}\big(\mathcal{C}^\infty(M, \bigwedge^k(M))\big) \subseteq \mathcal{C}^\infty(M, \bigwedge^{k+1}(M))$;
(2) $d(f) = df$ *(the ordinary differential of f) for every 0-form f;*
(3) *if ω is a k-form and μ is any differential form, then*

$$d(\omega \wedge \mu) = (d\omega) \wedge \mu + (-1)^k \omega \wedge d\mu; \quad \text{and}$$

(4) $d^2 = 0$.

Proof. Proofs of the existence and uniqueness of such a function can be found in [21] (Theorem 12.14), [32] (Chapter 1, Theorem 11.1), and [3] (Section 4.6). $\square$

Exercise 10.3.7. Let M be a 3-manifold, $\phi = (x, y, z) \colon U \to \mathbb{R}^3$ be a chart on M, and $f \colon U \to \mathbb{R}$ be a 0-form on U. Compute $d(f\,dy)$. (If f is a 0-form and ω is any differential form, it is conventional to write $f\omega$ for $f \wedge \omega$.)

Example 10.3.8. Let M be a 3-manifold and $\phi = (x, y, z) \colon U \to \mathbb{R}^3$ be a chart on M. Then $d\big(\cos(xy^2)\,dx \wedge dz\big) = 2xy\sin(xy^2)\,dx \wedge dy \wedge dz$.

Exercise 10.3.9. Let M be a 3-manifold and $\phi = (x, y, z) \colon U \to \mathbb{R}^3$ be a chart on M. Compute $d(x\,dy \wedge dz + y\,dz \wedge dx + z\,dx \wedge dy)$.

Exercise 10.3.10. Let M be a 3-manifold and $\phi = (x, y, z) \colon U \to \mathbb{R}^3$ be a chart on M. Compute $d[(3xz\,dx + xy^2\,dy) \wedge (x^2y\,dx - 6xy\,dz)]$.

Exercise 10.3.11. In beginning calculus texts some curious arguments are given for replacing the expression $dx\,dy$ in the integral $\iint_R f\,dx\,dy$ by $r\,dr\,d\theta$ when we change from rectangular to polar coordinates in the plane. Show that if we interpret $dx\,dy$ as the differential form $dx \wedge dy$, then this is a correct substitution. (Assume additionally that R is a region in the open first quadrant and that the integral of f over R exists.)

Exercise 10.3.12. Give an explanation similar to the one in the preceding exercise of the change in triple integrals from rectangular to spherical coordinates.

Exercise 10.3.13. Generalize the two preceding exercises.

Proposition 10.3.14. *If f is a 0-form and ω is a k-form on U, then* $*(f\omega) = f(*\omega)$.

Proposition 10.3.15. *If ω and μ are k-forms on U, then* $*(\omega + \mu) = *\omega + *\mu$.

Proposition 10.3.16. *If ω is a k-form on U, then* $* * \omega = (-1)^{k(n-k)}\omega$.

Notice that, in consequence of the preceding proposition, every k-form on a 3-manifold satisfies $* * \omega = \omega$.

Exercise 10.3.17. For real valued functions a, b, and c on U compute

(1) $*(a1)$,
(2) $*(a\,dx + b\,dy + c\,dz)$,
(3) $*(a\,dy \wedge dz + b\,dz \wedge dx + c\,dx \wedge dy)$, and
(4) $*a(dx \wedge dy \wedge dz)$.

10.4. Some Classical Vector Analysis

Definition 10.4.1. In beginning calculus we learn that the gradient of a smooth scalar field f on $\mathbb{R}^n$ can be represented at a point m as the vector $\left(\frac{\partial f}{\partial x^1}(m), \ldots, \frac{\partial f}{\partial x^n}(m)\right)$. For a smooth function f on the domain U of a chart $\phi = (x^1, \ldots, x^n)$ in an n-manifold we define the GRADIENT of f at a point m in U to be the vector *in the cotangent space* $T_m{}^*$ whose components with respect to the usual basis $\{dx^1{}_m, \ldots, dx^n{}_m\}$ for $T_m{}^*$ are just $\frac{\partial f}{\partial x^1}, \ldots, \frac{\partial f}{\partial x^n}$. We denote this vector by grad $f(m)$ or $\nabla f(m)$. Thus we make no distinction between the 1-forms grad f and df, since

$$\operatorname{grad} f = \frac{\partial f}{\partial x^1}\, dx^1 + \cdots + \frac{\partial f}{\partial x^n}\, dx^n = df$$

(see Proposition 9.5.36).

Definition 10.4.2. Let ω be a 1-form on the domain of a chart on a manifold. The CURL of ω, denoted by $\operatorname{curl}\omega$ or $\nabla \times \omega$ is defined by

$$\operatorname{curl}\omega = *\, d\omega.$$

(Notice that on a 3-manifold the curl of a 1-form is again a 1-form.)

Example 10.4.3. If $\omega = a\, dx + b\, dy + c\, dz$ is a 1-form on a 3-manifold, then

$$\operatorname{curl}\omega = \left(\frac{\partial c}{\partial y} - \frac{\partial b}{\partial z}\right) dx + \left(\frac{\partial a}{\partial z} - \frac{\partial c}{\partial x}\right) dy + \left(\frac{\partial b}{\partial x} - \frac{\partial a}{\partial y}\right) dz.$$

Remark 10.4.4. Some depraved souls who completely abandon all inhibitions concerning notation have been known to write

$$\operatorname{curl}\omega = \det \begin{bmatrix} dx & dy & dz \\ \frac{\partial}{\partial x} & \frac{\partial}{\partial y} & \frac{\partial}{\partial z} \\ a & b & c \end{bmatrix}.$$

Definition 10.4.5. Let ω be a 1-form on the domain of a chart on a manifold. The DIVERGENCE of ω, denoted by $\operatorname{div}\omega$ or $\nabla \cdot \omega$ is defined by

$$\operatorname{div}\omega = *\, d * \omega.$$

(Notice that on a 3-manifold the divergence of a 1-form is a 0-form; that is, a real valued function.)

Example 10.4.6. If $\omega = a\,dx + b\,dy + c\,dz$ is a 1-form on a 3-manifold, then

$$\operatorname{div}\omega = \frac{\partial a}{\partial x} + \frac{\partial b}{\partial y} + \frac{\partial c}{\partial z}.$$

Exercise 10.4.7. If f is a 0-form on the domain of a chart, prove (without using partial derivatives) that $\operatorname{curl}\operatorname{grad} f = 0$.

Exercise 10.4.8. If ω is a 1-form on the domain of a chart, prove (without using partial derivatives) that $\operatorname{div}\operatorname{curl}\omega = 0$.

Definition 10.4.9. Let ω and μ be 1-forms on the domain of a chart. Define the CROSS-PRODUCT of ω and μ, denoted by $\omega \times \mu$, by

$$\omega \times \mu = *(\omega \wedge \mu).$$

Example 10.4.10. If $\omega = a\,dx + b\,dy + c\,dz$ and $\mu = e\,dx + f\,dy + g\,dz$ are 1-forms on a 3-manifold, then

$$\omega \times \mu = (bg - cf)\,dx + (ce - ag)\,dy + (af - be)\,dz.$$

Remark 10.4.11. Occasionally as a memory aid some people write

$$\omega \times \mu = \det \begin{bmatrix} dx & dy & dz \\ a & b & c \\ e & f & g \end{bmatrix}.$$

Exercise 10.4.12. Suppose we wish to define the *dot product* $\langle \omega, \mu \rangle$ of two 1-forms $\omega = a\,dx + b\,dy + c\,dz$ and $\mu = e\,dx + f\,dy + g\,dz$ on a 3-manifold to be the 0-form $ae + bf + cg$. Rephrase this definition without mentioning the components of ω and μ.

Exercise 10.4.13. Suppose we wish to define the *triple scalar product* $[\omega, \mu, \eta]$ of the 1-forms $\omega = a\,dx + b\,dy + c\,dz$, $\mu = e\,dx + f\,dy + g\,dz$, and $\eta = j\,dx + k\,dy + l\,dz$ on a 3-manifold to be the 0-form $bgj - fcj + cek - agk + afl - bel$. Rephrase this definition without mentioning the components of ω, μ, and η.

10.5. Closed and Exact Forms

Definition 10.5.1. Let U be an open subset of a manifold and

$$\cdots \longrightarrow \bigwedge\nolimits^{k-1}(U) \xrightarrow{d_{k-1}} \bigwedge\nolimits^{k}(U) \xrightarrow{d_k} \bigwedge\nolimits^{k+1}(U) \longrightarrow \cdots$$

where d_{k-1} and d_k are exterior differentiation operators. Elements of $\ker d_k$ are called CLOSED k-forms and elements of $\operatorname{ran} d_{k-1}$ are EXACT k-forms. In other words, a k-form ω is CLOSED if $d\omega = 0$. It is EXACT if there exists a $(k-1)$-form η such that $\omega = d\eta$.

Proposition 10.5.2. *Every exact differential form is closed.*

Proposition 10.5.3. *If ω and μ are closed differential forms, so is $\omega \wedge \mu$.*

Proposition 10.5.4. *If ω is an exact form and μ is a closed form, then $\omega \wedge \mu$ is exact.*

Example 10.5.5. Let $\phi = (x, y, z) \colon U \to \mathbb{R}^3$ be a chart on a 3-manifold and $\omega = a\,dx + b\,dy + c\,dz$ be a 1-form on U. If ω is exact, then $\frac{\partial c}{\partial y} = \frac{\partial b}{\partial z}$, $\frac{\partial a}{\partial z} = \frac{\partial c}{\partial x}$, and $\frac{\partial b}{\partial x} = \frac{\partial a}{\partial y}$.

Exercise 10.5.6. Determine if each of the following 1-forms is exact in $\mathbb{R}^2$. If it is, specify the 0-form of which it is the differential.

(1) $ye^{xy}\,dx + xe^{xy}\,dy$;
(2) $x\sin y\,dx + x\cos y\,dy$; and
(3) $\left(\dfrac{\arctan y}{\sqrt{1-x^2}} + \dfrac{x}{y} + 3x^2 \right) dx + \left(\dfrac{\arcsin x}{1+y^2} - \dfrac{x^2}{2y^2} + e^y \right) dy.$

Exercise 10.5.7. Explain why solving the initial value problem

$$e^x \cos y + 2x - e^x (\sin y) y' = 0, \quad y(0) = \pi/3$$

is essentially the same thing as showing that the 1-form $(e^x \cos y + 2x)\,dx - e^x (\sin y)\,dy$ is exact. Do it.

10.6. Poincaré's Lemma

Notation 10.6.1 (for the entire section). Let $n \in \mathbb{N}$, U be a nonempty open convex subset of $\mathbb{R}^n$, and $x = (x^1, \ldots, x^n)$ be the identity map on U.

Whenever $1 \le k \le n$ and $\nu = b\,dx^{i_1} \wedge \cdots \wedge dx^{i_k}$ is a k-form on U, we define a 0-form g^ν on U by

$$g^\nu(x) := \int_0^1 b(tx) t^{k-1}\,dt \quad \text{if } \nu \ne 0 \quad \text{and} \quad g^0(x) = 0;$$

and we define $(k-1)$-forms μ^ν and $h(\nu)$ by

$$\mu^\nu := \sum_{j=1}^{k}(-1)^{j-1}x^{i_j}\,dx^{i_1}\wedge\cdots\wedge\widehat{dx^{i_j}}\wedge\cdots\wedge dx^{i_k}$$

and

$$h(\nu) := g^\nu\mu^\nu.$$

In the definition of μ^ν, the circumflex above the term dx^{i_j} indicates that the term is deleted. For example, if $k=3$, then

$$\mu^\nu = x^{i_1}\,dx^{i_2}\wedge dx^{i_3} - x^{i_2}\,dx^{i_1}\wedge dx^{i_3} + x^{i_3}\,dx^{i_1}\wedge dx^{i_2}.$$

For each k extend h to all of $\bigwedge^k(U)$ by requiring it to be linear. Thus

$$h^\rightarrow\left(\bigwedge^k(U)\right) \subseteq \bigwedge^{k-1}(U).$$

Theorem 10.6.2 (Poincaré's lemma). *If U is a nonempty open convex subset of $\mathbb{R}^n$ and $p \geq 1$, then every closed p-form on U is exact.*

Hint for proof. Let p be a fixed integer such that $1 \leq p \leq n$. Let $i_1,\ldots,i_p$ be distinct integers between 1 and n, and let a be a 0-form on U. Define

$$\beta := dx^{i_1}\wedge\cdots\wedge dx^{i_p},$$
$$\omega := a\beta,$$

and, for $1 \leq k \leq n$, define

$$\eta^k := a_k\,dx^k\wedge\beta.$$

(Here, $a_k = \frac{\partial a}{\partial x^k}$.)

Now, do the following.

(a) Show that $g^\omega{}_k(x) = \int_0^1 a_k(tx)t^p\,dt$ for $1 \leq k \leq n$ and $x \in U$.
(b) Show that $\mu^{\eta^k} = x^k\beta - dx^k\wedge\mu^\omega$ for $1 \leq k \leq n$.
(c) Compute $h(\eta^k)$ for $1 \leq k \leq n$.
(d) Show that $d\omega = \sum_{k=1}^n \eta^k$.
(e) Compute $h\,d\omega$.
(f) Show that $d(\mu^\omega) = p\,\beta$.
(g) Compute $d(h\omega)$.
(h) Compute $\dfrac{d}{dt}\left(t^p\,a(tx)\right)$.
(i) Show that $pg^\omega + \sum_{k=1}^n g^\omega{}_k x^k = a$.
(j) Show that $(dh + hd)(\omega) = \omega$.

Proposition 10.6.3. *If ω is a 1-form on a nonempty convex open subset U of $\mathbb{R}^3$ with* $\operatorname{curl}\omega = 0$, *then there exists a 0-form f on U such that* $\omega = \operatorname{grad} f$.

Exercise 10.6.4. *Use the proof of Poincaré's lemma to* find a 0-form on $\mathbb{R}^3$ whose gradient is

$$(2xyz^3 - y^2z)\,dx + (x^2z^3 - 2xyz)\,dy + (3x^2yz^2 - xy^2)\,dz.$$

Exercise 10.6.5. Consider the 1-form $\nu = e^z\,dx + x\,dy$ in $\mathbb{R}^3$ and let $\omega = d\nu$. *Use the proof of Poincaré's lemma to* find another 1-form $\eta = a\,dx + b\,dy + c\,dz$ such that $\omega = d\eta$. Explain carefully what happens at $z = 0$. Find $\dfrac{\partial^n a}{\partial z^n}(0,0,0)$ for every integer $n \geq 0$.

Proposition 10.6.6. *If ω is a 1-form on a nonempty convex open subset U of $\mathbb{R}^3$ with* $\operatorname{div}\omega = 0$, *then there exists a 1-form η on U such that* $\omega = \operatorname{curl}\eta$.

Exercise 10.6.7. Let $\omega = 2xyz\,dx + x^3z^2\,dy - yz^2\,dz$. Check that $\operatorname{div}\omega = 0$. *Use the proof of Poincaré's lemma to* find a 1-form η whose curl is ω.

Proposition 10.6.8. *Every smooth real valued function f on a nonempty convex open subset U of $\mathbb{R}^3$ is* $\operatorname{div}\eta$ *for some 1-form η on U.*

Exercise 10.6.9. In the proof of the preceding proposition, what needs to be changed if U lies in $\mathbb{R}^2$?

Exercise 10.6.10. *Use the proof of Poincaré's lemma to* find a 1-form η on $\mathbb{R}^3$ whose divergence is the function

$$f\colon (x,y,z) \mapsto xy - y^2z + xz^3.$$

Chapter 11

HOMOLOGY AND COHOMOLOGY

11.1. The de Rham Cohomology Group

Definition 11.1.1. Let M be an n-manifold. We denote by $Z^k(M)$ (or just Z^k) the vector space of all closed k-forms on M. The "Z" is for the German word "Zyklus", which means *cycle*. So in cohomological language closed forms are often called *cocycles*.

Also we denote by $B^k(M)$ (or just B^k) the vector space of exact k-forms on M. Since there are no differential forms of degree strictly less than 0, we take $B^0 = B^0(M) = \{0\}$. For convenience we also take $Z^k = \{0\}$ and $B^k = \{0\}$ whenever $k < 0$ or $k > n$. The letter "B" refers to the word "boundary". So exact forms in the context of cohomology are often called *coboundaries*.

It is a trivial consequence of Proposition 10.5.2 that $B^k(M)$ is a vector subspace of $Z^k(M)$. Thus it makes sense to define

$$H^k = H^k(M) := \frac{Z^k(M)}{B^k(M)}.$$

The quotient space $H^k(M)$ is the kth DE RHAM COHOMOLOGY GROUP of M. (Yes, even though it is a vector space, it is traditionally called a *group*.) The dimension of the vector space $H^k(M)$ is the kth BETTI NUMBER of the manifold M.

163

Another (obviously equivalent) way of phrasing the definition of the kth de Rham cohomology group is in terms of the maps

$$\cdots \longrightarrow \bigwedge^{k-1}(M) \xrightarrow{\ d_{k-1}\ } \bigwedge^{k}(M) \xrightarrow{\ d_k\ } \bigwedge^{k+1}(M) \longrightarrow \cdots$$

where d_{k-1} and d_k are exterior differentiation operators. Define

$$H^k(M) := \frac{\ker d_k}{\operatorname{ran} d_{k-1}}$$

for all k.

It is an interesting fact, but one that we shall not prove, that these cohomology groups are topological invariants. That is, if two manifolds M and N are homeomorphic, then $H^k(M)$ and $H^k(N)$ are isomorphic.

Example 11.1.2. If M is a connected manifold, then $H^0(M) \cong \mathbb{R}$.

Exercise 11.1.3. For U an open subset of $\mathbb{R}^n$ give a very clear description of $H^0(U)$ and explain why its dimension is the number of connected components of U. *Hint.* A function is said to be LOCALLY CONSTANT if it is constant in some neighborhood of each point in its domain.

Definition 11.1.4. Let $F \colon M \to N$ be a smooth function between smooth manifolds. For $k \geq 1$ define

$$\bigwedge^{k} F \colon \bigwedge^{k}(N) \to \bigwedge^{k}(M) \colon \omega \mapsto \left(\bigwedge^{k} F\right)(\omega)$$

where

$$\left(\left(\bigwedge^{k} F\right)(\omega)\right)_m (v^1, \ldots, v^k) = \omega_{F(m)}\big(dF_m(v^1), \ldots, dF_m(v^k)\big) \qquad (11.1)$$

for every $m \in M$ and $v^1, \ldots, v^k \in T_m$. Also define $\left(\left(\bigwedge^0 F\right)(\omega)\right)_m = \omega_{F(m)}$. We simplify the notation in (11.1) slightly

$$\left(\bigwedge^{k} F\right)(\omega)(v^1, \ldots, v^k) = \omega(dF(v^1), \ldots, dF(v^k)). \qquad (11.2)$$

Denote by F^* the map induced by the maps $\bigwedge^k F$ that takes the $\mathbb{Z}$-graded algebra $\bigwedge(N)$ to the $\mathbb{Z}$-graded algebra $\bigwedge(M)$.

Example 11.1.5. For each $k \in \mathbb{Z}^+$ the pair of maps $M \mapsto \bigwedge^k(M)$ and $F \mapsto \bigwedge^k(F)$ (as defined in 10.3.1 and 11.1.4) is a contravariant functor from the category of smooth manifolds and smooth maps to the category of vector spaces and linear maps.

Proposition 11.1.6. *If* $F\colon M \to N$ *is a smooth function between smooth manifolds,* $\omega \in \textstyle\bigwedge^{j}(N)$, *and* $\mu \in \textstyle\bigwedge^{k}(N)$, *then*

$$\left(\textstyle\bigwedge^{j+k} F\right)(\omega \wedge \mu) = \left(\textstyle\bigwedge^{j} F\right)(\omega) \wedge \left(\textstyle\bigwedge^{k} F\right)(\mu).$$

Exercise 11.1.7. In Example 11.1.5 you showed that $\textstyle\bigwedge^{k}$ was a functor for each k. What about $\textstyle\bigwedge$ itself? Is it a functor? Explain.

Proposition 11.1.8. *If* $F\colon M \to N$ *is a smooth function between smooth manifolds, then*

$$d \circ F^{*} = F^{*} \circ d.$$

Exercise 11.1.9. Let $V = \{0\}$ be the 0-dimensional Euclidean space. Compute the kth de Rham cohomology group $H^{k}(V)$ for all $k \in \mathbb{Z}$.

Exercise 11.1.10. Compute $H^{k}(\mathbb{R})$ for all $k \in \mathbb{Z}$.

Exercise 11.1.11. Let U be the union of m disjoint open intervals in $\mathbb{R}$. Compute $H^{k}(U)$ for all $k \in \mathbb{Z}$.

Exercise 11.1.12. Let U be an open subset of $\mathbb{R}^{n}$. For $[\omega] \in H^{j}(U)$ and $[\mu] \in H^{k}(U)$ define

$$[\omega][\mu] = [\omega \wedge \mu] \in H^{j+k}(U).$$

Explain why Proposition 10.5.3 is necessary for this definition to make sense. Prove also that this definition does not depend on the representatives chosen from the equivalence classes. Show that this definition makes $H^{*}(U) = \bigoplus_{k \in \mathbb{Z}} H^{k}(U)$ into a $\mathbb{Z}$-graded algebra. This is the DE RHAM COHOMOLOGY ALGEBRA of U.

Definition 11.1.13. Let $F\colon M \to N$ be a smooth function between smooth manifolds. For each integer k define

$$H^{k}(F)\colon H^{k}(N) \to H^{k}(M)\colon [\omega] \mapsto \left[\textstyle\bigwedge^{k}(F)(\omega)\right].$$

Denote by $H^{*}(F)$ the induced map that takes the $\mathbb{Z}$-graded algebra $H^{*}(N)$ into $H^{*}(M)$.

Example 11.1.14. With the definitions given in 11.1.12 and 11.1.13 H^{*} becomes a contravariant functor from the category of open subsets of $\mathbb{R}^{n}$ and smooth maps to the category of $\mathbb{Z}$-graded algebras and their homomorphisms.

11.2. Cochain Complexes

Definition 11.2.1. A sequence

$$\cdots \longrightarrow V_{k-1} \xrightarrow{\ d_{k-1}\ } V_k \xrightarrow{\ d_k\ } V_{k+1} \longrightarrow \cdots$$

of vector spaces and linear maps is a COCHAIN COMPLEX if $d_k \circ d_{k-1} = \mathbf{0}$ for all $k \in \mathbb{Z}$. Such a sequence may be denoted by (V^*, d) or just by V^*.

Definition 11.2.2. We generalize Definition 11.1.1 in the obvious fashion. If V^* is a cochain complex, then the kth COHOMOLOGY GROUP $H^k(V^*)$ is defined to be $\ker d_k / \operatorname{ran} d_{k-1}$. (As before, this "group" is actually a vector space.) In this context the elements of V_k are often called k-COCHAINS, elements of $\ker d_k$ are k-COCYCLES, elements of $\operatorname{ran} d_{k-1}$ are k-COBOUNDARIES, and d is the COBOUNDARY OPERATOR.

Definition 11.2.3. Let (V^*, d) and (W^*, δ) be cochain complexes. A COCHAIN MAP $G: V^* \to W^*$ is a sequence of linear maps $G_k: V_k \to W_k$ satisfying

$$\delta_k \circ G_k = G_{k+1} \circ d_k$$

for every $k \in \mathbb{Z}$. That is, the diagram

$$
\begin{array}{ccc}
\cdots \longrightarrow V_k & \xrightarrow{\ d_k\ } & V_{k+1} \longrightarrow \cdots \\
\downarrow{\scriptstyle G_k} & & \downarrow{\scriptstyle G_{k+1}} \\
\cdots \longrightarrow W_k & \xrightarrow[\ \delta_k\ ]{} & W_{k+1} \longrightarrow \cdots
\end{array}
$$

commutes.

Proposition 11.2.4. *Let $G: V^* \to W^*$ be a cochain map between cochain complexes. For each $k \in \mathbb{Z}$ define*

$$G_k^*: H^k(V^*) \to H^k(W^*): [v] \mapsto [G_k(v)]$$

whenever v is a cocycle in V_k. Then the maps G_k^* are well defined and linear.

Hint for proof. To prove that G_k^* is well-defined we need to show two things: that $G_k(v)$ is a cocycle in W_k and that the definition does not depend on the choice of representative v.

Definition 11.2.5. A sequence

$$0 \longrightarrow U^* \xrightarrow{F} V^* \xrightarrow{G} W^* \longrightarrow 0$$

of cochain complexes and cochain maps is (SHORT) EXACT if for every $k \in \mathbb{Z}$ the sequence

$$0 \longrightarrow U_k \xrightarrow{F_k} V_k \xrightarrow{G_k} W_k \longrightarrow 0$$

of vector spaces and linear maps is (short) exact.

Proposition 11.2.6. *If* $0 \longrightarrow U^* \xrightarrow{F} V^* \xrightarrow{G} W^* \longrightarrow 0$ *is a short exact sequence of cochain complexes, then*

$$H^k(U^*) \xrightarrow{F_k^*} H^k(V^*) \xrightarrow{G_k^*} H^k(W^*)$$

is exact at $H^k(V^*)$ *for every* $k \in \mathbb{Z}$.

Proposition 11.2.7. *A short exact sequence*

$$0 \longrightarrow U^* \xrightarrow{F} V^* \xrightarrow{G} W^* \longrightarrow 0$$

of cochain complexes induces a long exact sequence

$$\longrightarrow H^{k-1}(W^*) \xrightarrow{\eta_{k-1}} H^k(U^*) \xrightarrow{F_k^*} H^k(V^*) \xrightarrow{G_k^*} H^k(W^*)$$

$$\xrightarrow{\eta_k} H^{k+1}(U^*) \longrightarrow$$

Hint for proof. If w is a cocycle in W_k, then, since G_k is surjective, there exists $v \in V_k$ such that $w = G_k(v)$. It follows that $dv \in \ker G_{k+1} = \operatorname{ran} F_{k+1}$ so that $dv = F_{k+1}(u)$ for some $u \in U_{k+1}$. Let $\eta_k([w]) = [u]$.

11.3. Simplicial Homology

Definition 11.3.1. Let V be a vector space. Recall that a *linear combination* of a finite set $\{x_1, \ldots, x_n\}$ of vectors in V is a vector of the form $\sum_{k=1}^{n} \alpha_k x_k$ where $\alpha_1, \ldots, \alpha_n \in \mathbb{R}$. If $\alpha_1 = \alpha_2 = \cdots = \alpha_n = 0$, then the linear combination is *trivial*; if at least one α_k is different from zero, the linear combination is *nontrivial*. A linear combination $\sum_{k=1}^{n} \alpha_k x_k$ of the vectors $x_1, \ldots, x_n$ is a CONVEX COMBINATION if $\alpha_k \geq 0$ for each k $(1 \leq k \leq n)$ and if $\sum_{k=1}^{n} \alpha_k = 1$.

Definition 11.3.2. If a and b are vectors in the vector space V, then the CLOSED SEGMENT between a and b, denoted by $[a, b]$, is $\{(1-t)a + tb \colon 0 \le t \le 1\}$.

Caution 11.3.3. Notice that there is a slight conflict between this notation, when applied to the vector space $\mathbb{R}$ of real numbers, and the usual notation for closed intervals on the real line. In $\mathbb{R}$ the closed segment $[a, b]$ is the same as the closed interval $[a, b]$ provided that $a \le b$. If $a > b$, however, the closed segment $[a, b]$ is the same as the segment $[b, a]$, it contains all numbers c such that $b \le c \le a$, whereas the closed interval $[a, b]$ is empty.

Definition 11.3.4. A subset C of a vector space V is CONVEX if the closed segment $[a, b]$ is contained in C whenever a, $b \in C$.

Definition 11.3.5. Let A be a subset of a vector space V. The CONVEX HULL of A is the smallest convex subset of V that contain A.

Exercise 11.3.6. Show that Definition 11.3.5 makes sense by showing that the intersection of a family of convex subsets of a vector space is itself convex. Then show that a "constructive characterization" is equivalent; that is, prove that the convex hull of A is the set of all convex combinations of elements of A.

Definition 11.3.7. A set $S = \{v_0, v_1, \ldots, v_p\}$ of $p + 1$ vectors in a vector space V is CONVEX INDEPENDENT if the set $\{v_1 - v_0, v_2 - v_0, \ldots, v_p - v_0\}$ is linearly independent in V.

Definition 11.3.8. An AFFINE SUBSPACE of a vector space V is any translate of a linear subspace of V.

Example 11.3.9. The line whose equation is $y = 2x - 5$ is not a linear subspace of $\mathbb{R}^2$. But it is an affine subspace: it is the line determined by the equation $y = 2x$ (which *is* a linear subspace of $\mathbb{R}^2$) translated downwards parallel to the y-axis by 5 units.

Definition 11.3.10. Let $p \in \mathbb{Z}^+$. The closed convex hull of a convex independent set $S = \{v_0, \ldots, v_p\}$ of $p + 1$ vectors in some vector space is a CLOSED p-SIMPLEX. It is denoted by $[s]$ or by $[v_0, \ldots, v_p]$. The integer p is the DIMENSION of the simplex. The OPEN p-SIMPLEX determined by the set S is the set of all convex combinations $\sum_{k=0}^{p} \alpha_k v_k$ of elements of S where each $\alpha_k > 0$. The open simplex will be denoted by (s) or by $(v_0, \ldots, v_p)$.

We make the special convention that a single vector $\{v\}$ is both a closed and an open 0-simplex.

If $[s]$ is a simplex in $\mathbb{R}^n$ then the PLANE of $[s]$ is the affine subspace of $\mathbb{R}^n$ having the least dimension that contains $[s]$. It turns out that the open simplex (s) is the interior of $[s]$ in the plane of $[s]$.

Definition 11.3.11. Let $[s] = [v_0, \ldots, v_p]$ be a closed p-simplex in $\mathbb{R}^n$ and $\{j_0, \ldots, j_q\}$ be a nonempty subset of $\{0, 1, \ldots, p\}$. Then the closed q-simplex $[t] = [v_{j_0}, \ldots, v_{j_q}]$ is a CLOSED q-FACE of $[s]$. The corresponding open simplex (t) is an OPEN q-FACE of $[s]$. The 0-faces of a simplex are called the VERTICES of the simplex.

Note that distinct open faces of a closed simplex $[s]$ are disjoint and that the union of all the open faces of $[s]$ is $[s]$ itself.

Definition 11.3.12. Let $[s] = [v_0, \ldots, v_p]$ be a closed p-simplex in $\mathbb{R}^n$. We say that two orderings $(v_{i_0}, \ldots, v_{i_p})$ and $(v_{j_0}, \ldots, v_{j_p})$ of the vertices are EQUIVALENT if $(j_0, \ldots, j_p)$ is an even permutation of $(i_0, \ldots, i_p)$. (This *is* an equivalence relation.) For $p \geq 1$ there are exactly two equivalence classes; these are the ORIENTATIONS of $[s]$. An ORIENTED SIMPLEX is a simplex together with one of these orientations. The oriented simplex determined by the ordering $(v_0, \ldots, v_p)$ will be denoted by $\langle v_0, \ldots, v_p \rangle$. If, as above, $[s]$ is written as $[v_0, \ldots, v_p]$, then we may shorten $\langle v_0, \ldots, v_p \rangle$ to $\langle s \rangle$.

Of course, none of the preceding makes sense for 0-simplexes. We arbitrarily assign them two orientations, which we denote by $+$ and $-$. Thus $\langle s \rangle$ and $-\langle s \rangle$ have opposite orientations.

Definition 11.3.13. A finite collection K of open simplexes in $\mathbb{R}^n$ is a SIMPLICIAL COMPLEX if the following conditions are satisfied:

(1) if $(s) \in K$ and (t) is an open face of $[s]$, then $(t) \in K$; and
(2) if $(s), (t) \in K$ and $(s) \neq (t)$, then $(s) \cap (t) = \emptyset$.

The DIMENSION of a simplicial complex K, denoted by $\dim K$, is the maximum dimension of the simplexes constituting K. If $r \leq \dim K$, then the r-SKELETON of K, denoted by K^r, is the set of all open simplexes in K whose dimensions are no greater than r. The POLYHEDRON, $|K|$, of the complex K is the union of all the simplexes in K.

Definition 11.3.14. Let K be a simplicial complex in $\mathbb{R}^n$. For $0 \leq p \leq \dim K$ let $A_p(K)$ (or just A_p) denote the free vector space generated by the set of all oriented p-simplexes belonging to K. For $1 \leq p \leq \dim K$ let

$W_p(K)$ (or just W_p) be the subspace of A_p generated by all elements of the form

$$\langle v_0, v_1, v_2, \ldots, v_p \rangle + \langle v_1, v_0, v_2, \ldots, v_p \rangle$$

and let $C_p(K)$ (or just C_p) be the resulting quotient space A_p/W_p. For $p = 0$ let $C_p = A_p$ and for $p < 0$ or $p > \dim K$ let $C_p = \{0\}$. The elements of C_p are the p-CHAINS of K.

Notice that for any p we have

$$[\langle v_0, v_1, v_2, \ldots, v_p \rangle] = -[\langle v_1, v_0, v_2, \ldots, v_p \rangle].$$

To avoid cumbersome notation we will not distinguish between the p-chain $[\langle v_0, v_1, v_2, \ldots, v_p \rangle]$ and its representative $\langle v_0, v_1, v_2, \ldots, v_p \rangle$.

Definition 11.3.15. Let $\langle s \rangle = \langle v_0, v_1, \ldots, v_{p+1} \rangle$ be an oriented $(p+1)$-simplex. We define the BOUNDARY of $\langle s \rangle$, denoted by $\partial \langle s \rangle$, by

$$\partial \langle s \rangle = \sum_{k=0}^{p+1} (-1)^k \langle v_0, \ldots, \widehat{v_k}, \ldots, v_{p+1} \rangle.$$

(The caret above the v_k indicates that term is missing; so the boundary of a $(p+1)$-simplex is an alternating sum of p-simplexes.)

Definition 11.3.16. Let K be a simplicial complex in $\mathbb{R}^n$. For $1 \leq p \leq \dim K$ define

$$\partial_p = \partial \colon C_{p+1}(K) \to C_p(K) \colon$$

as follows. If $\sum a(s)\langle s \rangle$ is a p-chain in K, let

$$\partial\left(\sum a(s)\langle s \rangle \right) = \sum a(s)\partial \langle s \rangle.$$

For all other p let ∂_p be the zero map. The maps ∂_p are called BOUNDARY MAPS. Notice that each ∂_p is a linear map.

Proposition 11.3.17. *If K is a simplicial complex in $\mathbb{R}^n$, then $\partial^2 \colon C_{p+1}(K) \to C_{p-1}(K)$ is identically zero.*

Hint for proof. It suffices to prove this for generators $\langle v_0, \ldots, v_{p+1} \rangle$.

Definition 11.3.18. Let K be a simplicial complex in $\mathbb{R}^n$ and $0 \leq p \leq \dim K$. Define $Z_p(K) = Z_p$ to be the kernel of $\partial_{p-1} \colon C_p \to C_{p-1}$ and $B_p(K) = B_p$ to be the range of $\partial_p \colon C_{p+1} \to C_p$. The members of Z_p are p-CYCLES and the members of B_p are p-BOUNDARIES.

It is clear from Proposition 11.3.17 that B_p is a subspace of the vector space Z_p. Thus we may define $H_p(K) = H_p$ to be Z_p/B_p. It is the pth SIMPLICIAL HOMOLOGY GROUP of K. (And, of course, Z_p, B_p, and H_p are the trivial vector space whenever $p < 0$ or $p > \dim K$.)

Exercise 11.3.19. Let K be the topological boundary (that is, the 1-skeleton) of an oriented 2-simplex in $\mathbb{R}^2$. Compute $C_p(K)$, $Z_p(K)$, $B_p(K)$, and $H_p(K)$ for each p.

Exercise 11.3.20. What changes in Exercise 11.3.19 if K is taken to be the oriented 2-simplex itself?

Exercise 11.3.21. Let K be the simplicial complex in $\mathbb{R}^2$ comprising two triangular regions similarly oriented with a side in common. For all p compute $C_p(K)$, $Z_p(K)$, $B_p(K)$, and $H_p(K)$.

Definition 11.3.22. Let K be a simplicial complex. The number $\beta_p := \dim H_p(K)$ is the pth BETTI NUMBER of the complex K. And $\chi(K) := \sum_{p=0}^{\dim K} (-1)^p \beta_p$ is the EULER CHARACTERISTIC of K.

Proposition 11.3.23. *Let K be a simplicial complex. For $0 \leq p \leq \dim K$ let α_p be the number of p-simplexes in K. That is, $\alpha_p = \dim C_p(K)$. Then*

$$\chi(K) = \sum_{p=0}^{\dim K} (-1)^p \alpha_p.$$

11.4. Simplicial Cohomology

Definition 11.4.1. Let K be a simplicial complex. For each $p \in \mathbb{Z}$ let $C^p(K) = \big(C_p(K)\big)^*$. The elements of $C^p(K)$ are (SIMPLICIAL) p-COCHAINS. Then the adjoint $\partial_p{}^*$ of the boundary map

$$\partial_p \colon C_{p+1}(K) \to C_p(K)$$

is the linear map

$$\partial_p{}^* = \partial^* \colon C^p(K) \to C^{p+1}(K).$$

(Notice that $\partial^* \circ \partial^* = 0$.)

Also define

(1) $Z^p(K) := \ker \partial_p{}^*$;
(2) $B^p(K) := \operatorname{ran} \partial_{p-1}{}^*$; and
(3) $H^p(K) := Z^p(K)/B^p(K)$.

Elements of $Z^p(K)$ are (SIMPLICIAL) p-COCYCLES and elements of $B^p(K)$ are (SIMPLICIAL) p-COBOUNDARIES. The vector space $H^p(K)$ is the p^{TH} SIMPLICIAL COHOMOLOGY GROUP of K.

Proposition 11.4.2. *If K is a simplicial complex in $\mathbb{R}^n$, then $H^p(K) \cong \left(H_p(K)\right)^*$ for every integer p.*

Definition 11.4.3. Let $F\colon N \to M$ be a smooth injection between smooth manifolds. The pair (N, F) is a SMOOTH SUBMANIFOLD of M if dF_n is injective for every $n \in N$.

Definition 11.4.4. Let M be a smooth manifold, K be a simplicial complex in $\mathbb{R}^n$, and $h\colon [K] \to M$ be a homeomorphism. The triple (M, K, h) is a SMOOTHLY TRIANGULATED MANIFOLD if for every open simplex (s) in K the map $h\big|_{[s]}\colon [s] \to M$ has an extension $h_s\colon U \to M$ to a neighborhood U of $[s]$ lying in the plane of $[s]$ such that (U, h_s) is a smooth submanifold of M.

Theorem 11.4.5. *A smooth manifold can be triangulated if and only if it is compact.*

The proof of this theorem is tedious enough that very few textbook authors choose to include it in their texts. You can find a "simplified" proof in [5].

Theorem 11.4.6 (de Rham's theorem). *If (M, K, ϕ) is a smoothly triangulated manifold, then*

$$H^p(M) \cong H^p(K)$$

for every $p \in \mathbb{Z}$.

Proof. See [17], Chapter IV, Theorem 3.1; [21], Theorem 16.12; [30], pages 164–173; and [33], Theorem 4.17. □

Proposition 11.4.7 (pullbacks of differential forms). *Let $F\colon M \to N$ be a smooth mapping between smooth manifolds. Then there exists an algebra homomorphism $F^*\colon \bigwedge(N) \to \bigwedge(M)$, called the PULLBACK associated with F that satisfies the following conditions:*

(1) *F^* maps $\bigwedge^p(N)$ into $\bigwedge^p(M)$ for each p;*
(2) *$F^*(g) = g \circ F$ for each 0-form g on N; and*

(3) $(F^*\mu)_m(v) = \mu_{F(m)}(dF_m(v))$ *for every 1-form μ on N, every $m \in M$, and every $v \in T_m$.*

Proposition 11.4.8. *If $F\colon M \to N$ is a smooth map between n-manifolds, then F^* is a cochain map from the cochain complex $(\bigwedge^*(N), d)$ to the cochain complex $(\bigwedge^*(M), d)$. That is, the diagram*

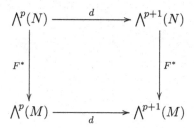

commutes for every $p \in \mathbb{Z}$.

Chapter 12

STOKES' THEOREM

12.1. Integration of Differential Forms

Definition 12.1.1. Let $\langle s \rangle$ be an oriented p-simplex in $\mathbb{R}^n$ (where $1 \leq p \leq n$) and μ be a p-form defined on a set U that is open in the plane of $\langle s \rangle$ and contains $[s]$. If $\langle s \rangle = \langle v_0, \ldots, v_p \rangle$ take $(v_1 - v_0, \ldots, v_p - v_0)$ to be an ordered basis for the plane of $\langle s \rangle$ and let $x^1, \ldots, x^p$ be the coordinate projection functions relative to this ordered basis; that is, if $a = \sum_{k=1}^{p} a_k(v_k - v_0) \in U$, then $x^j(a) = a_j$ for $1 \leq j \leq p$. Then $\phi = (x^1, \ldots, x^p) \colon U \to \mathbb{R}^p$ is a chart on U; so there exists a smooth function g on U such that $\mu = g \, dx^1 \wedge \cdots \wedge dx^p$. Define

$$\int_{\langle s \rangle} \mu = \int_{[s]} g \, dx^1 \ldots dx^p$$

where the right hand side is an ordinary Riemann integral. If $\langle v_0 \rangle$ is a 0-simplex, we make a special definition

$$\int_{\langle v_0 \rangle} f = f(v_0)$$

for every 0-form f.

Extend the preceding definition to p-chains by requiring the integral to be linear as a function of simplexes; that is, if $c = \sum a_s \langle s \rangle$ is a p-chain (in some simplicial complex) and μ is a p-form, define

$$\int_c \mu = \sum a(s) \int_{\langle s \rangle} \mu.$$

Definition 12.1.2. For a smoothly triangulated manifold (M, K, h) we define a map

$$\int_p : \bigwedge^p(M)' \to C^p(K)$$

as follows. If ω is a p-form on M, then $\int_p \omega$ is to be a linear functional on $C_p(K)$; that is, a member of $C^p(K) = \left(C_p(K) \right)^*$. In order to define a linear functional on $C_p(K)$ it suffices to specify its values on the basis vectors of $C_p(K)$; that is, on the oriented p-simplexes $\langle s \rangle$ that constitute $C_p(K)$. Let $h_s : U \to M$ be an extension of $h\big|_{[s]}$ to an open set U in the plane of $\langle s \rangle$. Then h_s^* pulls back p-forms on M to p-forms on U so that $h_s^*(\omega) \in \bigwedge^p(U)$. Define

$$\left(\int_p \omega \right) \langle s \rangle := \int_{\langle s \rangle} h_s^*(\omega).$$

Exercise 12.1.3. Let V be an open subset of $\mathbb{R}^n$, $F \colon V \to \mathbb{R}^n$, and $c \colon [t_0, t_1] \to V$ be a smooth curve in V. Let $C = \operatorname{ran} c$. It is conventional to define the "integral of the tangential component of F over C", often denoted by $\int_C F_T$, by the formula

$$\int_C F_T = \int_{t_0}^{t_1} \langle F \circ c, Dc \rangle = \int_{t_0}^{t_1} \langle F(c(t)), c'(t) \rangle \, dt. \tag{12.1}$$

The "tangential component of F", written F_T may be regarded as the 1-form $\sum_{k=1}^n F^k \, dx^k$.

Make sense of the preceding definition in terms of the definition of the integral of 1-forms over a smoothly triangulated manifold. For simplicity take $n = 2$. *Hint.* Suppose we have the following:

(1) $\langle t_0, t_1 \rangle$ (with $t_0 < t_1$) is an oriented 1-simplex in $\mathbb{R}$;
(2) V is an open subset of $\mathbb{R}^2$;
(3) $c \colon J \to V$ is an injective smooth curve in V, where J is an open interval containing $[t_0, t_1]$; and
(4) $\omega = a \, dx + b \, dy$ is a smooth 1-form on V.

First show that

$$\left(c^*(dx) \right)(t) = Dc^1(t)$$

for $t_0 \leq t \leq t_1$. (We drop the notational distinction between c and its extension c_s to J. Since the tangent space T_t is one-dimensional for every t, we identify T_t with $\mathbb{R}$. Choose v (in (3) of Proposition 11.4.7) to be the usual basis vector in $\mathbb{R}$, the number 1.)

Show in a similar fashion that

$$\left(c^*(dy)\right)(t) = Dc^2(t).$$

Then write an expression for $\left(c^*(\omega)\right)(t)$. Finally conclude that $\left(\int_1 \omega\right)(\langle t_0, t_1 \rangle)$ is indeed equal to $\int_{t_0}^{t_1} \langle (a, b) \circ c, Dc \rangle$ as claimed in 12.1.

Exercise 12.1.4. Let $\mathbb{S}^1$ be the unit circle in $\mathbb{R}^2$ oriented counterclockwise and let $\mathbf{F}$ be the vector field defined by $\mathbf{F}(x, y) = (2x^3 - y^3)\mathbf{i} + (x^3 + y^3)\mathbf{j}$. Use your work in Exercise 12.1.3 to calculate $\int_{\mathbb{S}^1} F_T$. *Hint.* You may use without proof two facts: (1) the integral does not depend on the parametrization (triangulation) of the curve, and (2) the results of Exercise 12.1.3 hold also for simple closed curves in $\mathbb{R}^2$; that is, for curves $c \colon [t_0, t_1] \to \mathbb{R}^2$ that are injective on the open interval (t_0, t_1) but that also satisfy $c(t_0) = c(t_1)$.

Notation 12.1.5. Let $\mathbb{H}^n = \{x \in \mathbb{R}^n \colon x_n \geq 0\}$. This is the UPPER HALF-SPACE of $\mathbb{R}^n$.

Definition 12.1.6. A n-MANIFOLD WITH BOUNDARY is defined in the same way as an n-manifold except that the range of a chart is assumed to be an open subset of $\mathbb{H}^n$.

The INTERIOR of $\mathbb{H}^n$, denoted by int $\mathbb{H}^n$, is defined to be $\{x \in \mathbb{R}^n \colon x_n > 0\}$. (Notice that this is the interior of $\mathbb{H}^n$ regarded as a subset of $\mathbb{R}^n$ — **not** of $\mathbb{H}^n$.) The BOUNDARY of $\mathbb{H}^n$, denoted by $\partial \mathbb{H}^n$, is defined to be $\{x \in \mathbb{R}^n \colon x_n = 0\}$.

If M is an n-manifold with boundary, a point $m \in M$ belongs to the INTERIOR of M (denoted by int M) if $\phi(m) \in$ int $\mathbb{H}^n$ for some chart ϕ. And it belongs to the BOUNDARY of M (denoted by ∂M) if $\phi(m) \in \partial \mathbb{H}^n$ for some chart ϕ.

Theorem 12.1.7. *Let M and N be a smooth n-manifolds with boundary and $F \colon M \to N$ be a smooth diffeomorphism. Then both int M and ∂M are smooth manifolds (without boundary). The interior of M has dimension n and the boundary of M has dimension $n - 1$. The mapping F induces smooth diffeomorphisms int $F \colon$ int $M \to$ int N and $\partial F \colon \partial M \to \partial N$.*

Proof. Consult the marvelous text [2], Proposition 7.2.6. $\qquad\square$

Exercise 12.1.8. Let V be an open subset of $\mathbb{R}^3$, $\mathbf{F} \colon V \to \mathbb{R}^3$ be a smooth vector field, and (S, K, h) be a smoothly triangulated 2-manifold such that $S \subseteq V$. It is conventional to define the "normal component of $\mathbf{F}$ over S ", often denoted by $\iint_S \mathbf{F}_N$, by the formula

$$\iint_S \mathbf{F}_N = \iint_K \langle \mathbf{F} \circ h, n \rangle$$

where $n = h_1 \times h_2$. (Notation: h_k is the kth partial derivative of h.)

Make sense of the preceding definition in terms of the definition of the integral of 2-forms over a smoothly triangulated manifold (with or without boundary). In particular, suppose that $\mathbf{F} = a\,\mathbf{i} + b\,\mathbf{j} + c\,\mathbf{k}$ (where a, b, and c are smooth functions) and let $\omega = a\,dy \wedge dz + b\,dz \wedge dx + c\,dx \wedge dy$. This 2-form is conventionally called the "normal component of $\mathbf{F}$" and is denoted by $\mathbf{F}_N$. Notice that $\mathbf{F}_N$ is just $*\mu$ where μ is the 1-form associated with the vector field $\mathbf{F}$. *Hint.* Proceed as follows.

(a) Show that the vector $n(u, v)$ is perpendicular to the surface S at $h(u, v)$ for each (u, v) in $[K]$ by showing that it is perpendicular to $D(h \circ c)(0)$ whenever c is a smooth curve in $[K]$ such that $c(0) = (u, v)$.

(b) Let u and v (in that order) be the coordinates in the plane of $[K]$ and x, y, and z (in that order) be the coordinates in $\mathbb{R}^3$. Show that $h^*(dx) = h_1^1\,du + h_2^1\,dv$. Also compute $h^*(dy)$ and $h^*(dz)$.

Remark. If at each point in $[K]$ we identify the tangent plane to $\mathbb{R}^2$ with $\mathbb{R}^2$ itself and if we use conventional notation, the "v" that appears in (3) of Proposition 11.4.7 is just not written. One keeps in mind that the components of h and all the differential forms are functions on (a neighborhood of) $[K]$.

(c) Now find $h^*(\omega)$. (Recall that $\omega = \mathbf{F}_N$ is defined above.)

(d) Show for each simplex (s) in K that

$$\left(\int_2 \omega \right)(\langle s \rangle) = \iint_{[s]} \langle \mathbf{F} \circ h, n \rangle.$$

(e) Finally show that if $\langle s_1 \rangle, \ldots, \langle s_n \rangle$ are the oriented 2-simplexes of K and $c = \sum_{k=1}^n \langle s_k \rangle$, then

$$\left(\int_2 \omega \right)(c) = \iint_{[K]} \langle \mathbf{F} \circ h, n \rangle.$$

Exercise 12.1.9. Let $\mathbf{F}(x, y, z) = xz\,\mathbf{i} + yz\,\mathbf{j}$ and H be the hemisphere of $x^2 + y^2 + z^2 = 4$ for which $z \geq 0$. Use Exercise 12.1.8 to find $\iint_H \mathbf{F}_N$.

12.2. Generalized Stokes' Theorem

Theorem 12.2.1 (Generalized Stokes' theorem). *Suppose that* (M, K, h) *is an oriented smoothly triangulated manifold with boundary. Then the integration operator* $\int = \left(\int_p\right)_{p \in \mathbb{Z}}$ *is a cochain map from the cochain complex* $(\bigwedge^*(M), d)$ *to the cochain complex* $(C^*(K), \partial^*)$.

Proof. This is an important and standard theorem, which appears in many versions and with many different proofs. See, for example, [2], Theorem 7.2.6; [20], Chapter XVII, Theorem 2.1; [26], Theorem 10.8; or [33], Theorems 4.7 and 4.9.

Recall that when we say in *Stokes' Theorem* that the integration operator is a cochain map, we are saying that the following diagram commutes.

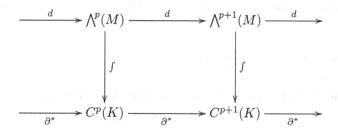

Thus if ω is a p-form on M and $\langle s \rangle$ is an oriented $(p+1)$-simplex belonging to K, then we must have

$$\left(\int_{p+1} d\omega\right)(\langle s \rangle) = \left(\partial^*\left(\int_p \omega\right)\right)(\langle s \rangle). \tag{12.2}$$

This last Equation (12.2) can be written in terms of integration over oriented simplexes:

$$\int_{\langle s \rangle} d(h_s{}^*\omega) = \int_{\partial\langle s \rangle} h_s{}^*\omega. \tag{12.3}$$

In more conventional notation all mention of the triangulating simplicial complex K and of the map h is suppressed. This is justified by the fact that it can be shown that the value of the integral is independent of the particular triangulation used. Then when the equations of the form (12.3) are added

over all the $(p + 1)$-simplexes comprising K we arrive at a particularly
simple formulation of (the conclusion of) *Stokes' Theorem*

$$\int_M d\omega = \int_{\partial M} \omega. \tag{12.4}$$

$\square$

One particularly important topic that has been glossed over in the preceding
is a discussion of orientable manifolds (those that possess nowhere vanishing
volume forms), their orientations, and the manner in which an orientation
of a manifold with boundary induces an orientation on its boundary. One
of many places where you can find a careful development of this material
is in Sections 6.5 and 7.2 of [2].

Theorem 12.2.2. *Let ω be a 1-form on a connected open subset U of $\mathbb{R}^2$.
Then ω is exact on U if and only if $\int_C \omega = 0$ for every simple closed curve
in U.*

Proof. See [9], Chapter 2, Proposition 1. $\square$

Example 12.2.3. Let $\omega = -\frac{y\,dx}{x^2+y^2} + \frac{x\,dy}{x^2+y^2}$. On the region $\mathbb{R}^2 \setminus \{(0,0)\}$
the 1-form ω is closed but not exact.

Exercise 12.2.4. What classical theorem do we get from the version of
Stokes' Theorem given by Equation (12.4) in the special case that M is a
flat 1-manifold (with boundary) in $\mathbb{R}$ and ω is a 0-form defined on some
open set in $\mathbb{R}$ that contains M? Explain.

Exercise 12.2.5. What classical theorem do we get from the version of
Stokes' Theorem given by Equation (12.4) in the special case that M is a
(not necessarily flat) 1-manifold (with boundary) in $\mathbb{R}^3$ and ω is a 0-form
defined on some open subset of $\mathbb{R}^3$ that contains M? Explain.

Exercise 12.2.6. What classical theorem do we get from the version of
Stokes' Theorem given by Equation (12.4) in the special case that M is a
flat 2-manifold (with boundary) in $\mathbb{R}^2$ and ω is the 1-form associated with a
vector field $\mathbf{F}\colon U \to \mathbb{R}^2$ defined on an open subset U of $\mathbb{R}^2$ that contains M?
Explain.

Exercise 12.2.7. Use Exercise 12.2.6 to compute $\int_{\mathbb{S}^1}(2x^3 - y^3)\,dx +
(x^3 + y^3)\,dy$ (where $\mathbb{S}^1$ is the unit circle oriented counterclockwise).

Exercise 12.2.8. Let $\mathbf{F}(x, y) = (-y, x)$ and let C_a and C_b be the circles centered at the origin with radii a and b, respectively, where $a < b$. Suppose that C_a is oriented clockwise and C_b is oriented counterclockwise. Find

$$\int_{C_a} \mathbf{F} \cdot d\mathbf{r} + \int_{C_b} \mathbf{F} \cdot d\mathbf{r}.$$

Exercise 12.2.9. What classical theorem do we get from the version of *Stokes' Theorem* given by Equation (12.4) in the special case that M is a (not necessarily flat) 2-manifold (with boundary) in $\mathbb{R}^3$ and ω is the 1-form associated with a vector field $\mathbf{F} \colon U \to \mathbb{R}^3$ defined on an open subset U of $\mathbb{R}^3$ that contains M? Explain.

Exercise 12.2.10. What classical theorem do we get from the version of *Stokes' Theorem* given by Equation (12.4) in the special case that M is a (flat) 3-manifold (with boundary) in $\mathbb{R}^3$ and $\omega = *\mu$ where μ is the 1-form associated with a vector field $\mathbf{F} \colon U \to \mathbb{R}^3$ defined on an open subset U of $\mathbb{R}^3$ that contains M? Explain.

Exercise 12.2.11. Your good friend Fred R. Dimm calls you on his cell phone seeking help with a math problem. He says that he wants to evaluate the integral of the normal component of the vector field on $\mathbb{R}^3$ whose coordinate functions are x, y, and z (in that order) over the surface of a cube whose edges have length 4. Fred is concerned that he's not sure of the coordinates of the vertices of the cube. How would you explain to Fred (over the phone) that it doesn't matter where the cube is located and that it is entirely obvious that the value of the surface integral he is interested in is 192?

Chapter 13

GEOMETRIC ALGEBRA

13.1. Geometric Plane Algebra

In this chapter we look at an interesting algebraic approach to plane geometry, one that keeps track of orientation of plane figures as well as such standard items as magnitudes, direction, perpendicularity, and angle. The object we start with is the Grassmann algebra $\bigwedge(\mathbb{R}^2)$ on which we define a new multiplication, called the *Clifford product*, under which it will become, in the language of the next chapter, an example of a *Clifford algebra*.

Notation 13.1.1. We make a minor notational change. Heretofore the standard basis vectors for $\mathbb{R}^n$ were denoted by $e^1, e^2, \ldots, e^n$. For our present purposes we will list them as $e_1, e_2, \ldots, e_n$. This allows us to use superscripts for powers.

Definition 13.1.2. We start with the inner product space $\mathbb{R}^2$. On the corresponding Grassmann algebra $\bigwedge(\mathbb{R}^2) = \bigoplus_{k=0}^{2} \bigwedge^k(\mathbb{R}^2)$ we say that an element homogeneous of degree 0 (that is, belonging to $\bigwedge^0(\mathbb{R}^2)$) is a SCALAR (or real number). An element homogeneous of degree 1 (that is, belonging to $\bigwedge^1(\mathbb{R}^2)$) we will call a VECTOR (or a 1-VECTOR). And an element homogeneous of degree 2 (that is, belonging to $\bigwedge^2(\mathbb{R}^2)$) will be called a BIVECTOR (or a 2-BLADE, or a PSEUDOSCALAR). We will shorten the name of the bivector $e_1 \wedge e_2$ to e_{12}. Similarly we write e_{21} for the bivector $e_2 \wedge e_1$. Notice that $\{1, e_1, e_2, e_{12}\}$ is a basis for the 4-dimensional Grassmann algebra $\bigwedge(\mathbb{R}^2)$, and that $e_{21} = -e_{12}$. Notice also that since $\bigwedge^2(\mathbb{R}^2)$ is 1-dimensional, every bivector is a scalar multiple of e_{12}.

We now introduce a new multiplication, called CLIFFORD MULTIPLICA-TION (or GEOMETRIC MULTIPLICATION) into $\bigwedge(\mathbb{R}^2)$ as follows: if v and w are 1-vectors in $\bigwedge(\mathbb{R}^2)$, then their CLIFFORD PRODUCT vw is defined by

$$vw := \langle v, w \rangle + v \wedge w. \tag{13.1}$$

To make $\bigwedge(\mathbb{R}^2)$ into a unital algebra, we will extend this new multiplication to all elements of the space. The Clifford product of a scalar and an arbitrary element of $\bigwedge(\mathbb{R}^2)$ is the same as it is in the Grassmann algebra $\bigwedge(\mathbb{R}^2)$. We must also specify the Clifford product of a vector and a bivector. Since we require the distributive law to hold for Clifford multiplication we need only specify the product $e_i e_{12}$ and $e_{12} e_i$ for $i = 1$ and 2. If we wish Clifford multiplication to be associative there is only one way to do this (see Exercise 13.1.7): set $e_1 e_{12} = e_2$, $e_2 e_{12} = -e_1$, $e_{12} e_1 = -e_2$, and $e_{12} e_2 = e_1$. Similarly, if we are to have associativity, the product of bivectors must be determined by $e_{12} e_{12} = -1$.

Caution 13.1.3. Now we have three different "multiplications" to consider: the *inner* (or *dot*) product $\langle v, w \rangle$, the wedge (or *exterior*, or *outer*) product $v \wedge w$, and the *Clifford* (or *geometric*) product vw. Equation (13.1) holds only for *vectors*. Do *not* use it for arbitrary members of $\bigwedge(\mathbb{R}^2)$. (There is, of course, a possibility of confusion: Since $\bigwedge(\mathbb{R}^2)$ is indeed a vector space, any of its elements may in general usage be called a "vector". On the other hand, we have divided the homogeneous elements of $\bigwedge(\mathbb{R}^2)$ into three disjoint classes: scalars, vectors, and bivectors. When we wish to emphasize that we are using the word "vector" in the latter sense, we call it a 1-vector.

Notation 13.1.4. The object we have just defined turns out to be a unital algebra and is an example of a *Clifford algebra*. In the next chapter we will explain why we denote this particular algebra — the vector space $\bigwedge(\mathbb{R}^2)$ together with Clifford multiplication — by $\mathrm{Cl}(2,0)$.

Proposition 13.1.5. *If v is a 1-vector in $\mathrm{Cl}(2,0)$, then $v^2 = \|v\|^2$.* (Here, v^2 means vv.)

Corollary 13.1.6. *Every nonzero 1-vector v in $\mathrm{Cl}(2,0)$ has an inverse (with respect to Clifford multiplication). As usual, the inverse of v is denoted by v^{-1}.*

Exercise 13.1.7. Justify the claims made in the last two sentences of Definition 13.1.2.

The magnitude (or norm) of a 1-vector is just its length. We will also assign a MAGNITUDE (or NORM) to bivectors. The motivation for the following definition is that we wish a bivector $v \wedge w$ to represent an equivalence class of directed regions in the plane, two such regions being equivalent if they have the same area and the same orientation (positive = counterclockwise or negative = clockwise). So we will take the magnitude of $v \wedge w$ to be the area of the parallelogram generated by v and w.

Definition 13.1.8. Let v and w be 1-vectors in $\mathrm{Cl}(2,0)$. Define $\|v \wedge w\| := \|v\| \|w\| \sin \theta$ where θ is the angle between v and w $(0 \le \theta \le \pi)$.

Proposition 13.1.9. *If v and w are 1-vectors in $\mathrm{Cl}(2,0)$, then $\|v \wedge w\|$ is the area of the parallelogram generated by v and w.*

Proposition 13.1.10. *If v and w are 1-vectors in $\mathrm{Cl}(2,0)$, then $v \wedge w = \|v \wedge w\| e_{12} = \det(v, w) e_{12}$.*

Exercise 13.1.11. Suppose you know how to find the Clifford product of any two elements of $\mathrm{Cl}(2,0)$. Explain how to use Equation (13.1) to recapture formulas defining the inner product and the wedge product of two 1-vectors in $\mathrm{Cl}(2,0)$.

A vector in the plane is usually regarded as an equivalence class of directed segments. (Two directed segments are taken to be equivalent if they lie on parallel lines, have the same length, and point in the same direction.) Each such equivalence class of directed segments has a *standard representative*, the one whose tail is at the origin. Two standard representatives are parallel if one is a nonzero multiple of the other. By a common abuse of language, where we conflate the notions of directed segments and vectors, we say that two nonzero vectors in the plane are parallel if one is a scalar multiple of the other.

Proposition 13.1.12. *Two nonzero vectors in* $\mathrm{Cl}(2,0)$ *are parallel if and only if their Clifford product commutes.*

Definition 13.1.13. Two elements, a and b, in a semigroup (or ring, or algebra) ANTICOMMUTE if $ba = -ab$.

Example 13.1.14. The bivector e_{12} anticommutes with all vectors in $\mathrm{Cl}(2,0)$.

Proposition 13.1.15. *Two vectors in* $\mathrm{Cl}(2,0)$ *are perpendicular if and only if their Clifford product anticommutes.*

Let v and w be nonzero vectors in $\mathrm{Cl}(2,0)$. We can write v as the sum of two vectors $v_{\parallel}$ and $v_{\perp}$, where $v_{\parallel}$ is parallel to w and $v_{\perp}$ is perpendicular to w. The vector $v_{\parallel}$ is the PARALLEL COMPONENT of v and $v_{\perp}$ is the PERPENDICULAR COMPONENT of v (with respect to w).

Proposition 13.1.16. *Let v and w be nonzero vectors in* $\mathrm{Cl}(2,0)$. *Then* $v_{\parallel} = \langle v, w \rangle w^{-1}$ *and* $v_{\perp} = (v \wedge w) w^{-1}$.

Definition 13.1.17. If v and w are nonzero 1-vectors in $\mathrm{Cl}(0,2)$, the REFLECTION of v across w is the map $v = v_{\parallel} + v_{\perp} \mapsto v_{\parallel} - v_{\perp}$.

Proposition 13.1.18. *If v and w are nonzero 1-vectors in* $\mathrm{Cl}(0,2)$, *then the reflection $v' = v_{\parallel} - v_{\perp}$ of v across w is given by $v' = wvw^{-1}$.*

Although $\mathrm{Cl}(2,0)$ is a real algebra, it contains a field isomorphic to the complex numbers. We start with the observation that $e_{12}{}^2 = -1$. This prompts us to give a new name to e_{12}. We will call it $\hat{\imath}$. In the space $\bigwedge^0(\mathbb{R}^2) \oplus \bigwedge^2(\mathbb{R}^2)$, of scalars plus pseudoscalars, this element serves the same role as the complex number i does in the complex plane $\mathbb{C}$.

Definition 13.1.19. If $\zeta := a + b\hat{\imath}$ is the sum of a scalar and a pseudoscalar, we define its CONJUGATE $\bar{\zeta}$ as $a - b\hat{\imath}$ and its ABSOLUTE VALUE $|\zeta|$ as $\sqrt{a^2 + b^2}$.

Proposition 13.1.20. *If $\zeta \in \bigwedge^0(\mathbb{R}^2) \oplus \bigwedge^2(\mathbb{R}^2)$, then $\zeta\bar{\zeta} = |\zeta|^2$ and, if ζ is not zero, it is invertible with $\zeta^{-1} = \bar{\zeta}|\zeta|^{-2}$.*

Proposition 13.1.21. *The space $\bigwedge^0(\mathbb{R}^2) \oplus \bigwedge^2(\mathbb{R}^2)$ (with Clifford multiplication) is a field and is isomorphic to the field $\mathbb{C}$ of complex numbers.*

Complex numbers serve two purposes in the plane. They implement rotations via their representation in polar form $ze^{i\theta}$ and, as points, they represent vectors. Recall that in the Clifford algebra $Cl(2,0)$ we have $\bigwedge^1(\mathbb{R}^2) = \mathbb{R}^2$. So here vectors live in $\bigwedge^1(\mathbb{R}^2)$ while (the analogs of) complex numbers live in $\bigwedge^0(\mathbb{R}^2) \oplus \bigwedge^2(\mathbb{R}^2)$. Since these are both two-dimensional vector spaces, they are isomorphic. It turns out that there exists a simple, natural, and very useful isomorphism between these spaces: left multiplication by e_1.

Notation 13.1.22. In the remainder of this section we will shorten $\bigwedge(\mathbb{R}^2)$ to $\bigwedge$ and $\bigwedge^k(\mathbb{R}^2)$ to $\bigwedge^k$ for $k = 0$, 1, and 2.

Proposition 13.1.23. *The map from $\bigwedge^1$ to $\bigwedge^0 \oplus \bigwedge^2$ defined by $v \mapsto \hat{v} := e_1 v$ is a vector space isomorphism. The mapping is also an isometry in the sense that $|\hat{v}| = \|v\|$. The mapping is its own inverse; that is, $e_1\hat{v} = v$.*

Definition 13.1.24. For $\phi \in \mathbb{R}$ define $e^{\phi \hat{\iota}}$ by the usual power series

$$e^{\phi \hat{\iota}} := \sum_{k=0}^{\infty} \frac{(\phi \hat{\iota})^k}{k!} = \cos\phi + (\sin\phi)\,\hat{\iota}.$$

Exercise 13.1.25. Discuss the convergence problems (or lack thereof) that arise in the preceding definition.

Next we develop a formula for the positive (counterclockwise) rotation of a nonzero vector v through angle of ϕ radians. Let v' be the vector in its rotated position. First move these 1-vectors to their corresponding points in the ($Cl(2,0)$ analog of the) complex plane. Regarding these two points, $\zeta = e_1 v$, $\zeta' = e_1 v' \in \bigwedge^0 \oplus \bigwedge^2$ as complex numbers (*via* Proposition 13.1.21), we see that $\zeta' = e^{\phi\hat{\iota}}\zeta$. Then move ζ and ζ' back to $\bigwedge^1$ to obtain the formula in the following proposition.

Proposition 13.1.26. *Let v be a nonzero 1-vector in $Cl(2,0)$. After being rotated by ϕ radians in the positive direction it becomes the vector v'. Then*

$$v' = e^{-\phi\hat{\iota}}v = ve^{\phi\hat{\iota}}.$$

In Definition 13.1.2 and Exercise 13.1.7 we established that a necessary condition for $Cl(2,0)$ under Clifford multiplication to be an algebra, is that certain relations involving the basis elements, 1, e_1, e_2, and e_{12} hold. This doesn't prove, however, that if these relations are satisfied, then $Cl(2,0)$ under Clifford multiplication is an algebra. This lacuna is not hard to fill.

Proposition 13.1.27. *The Clifford algebra* $\mathrm{Cl}(2,0)$ *is, in fact, a unital algebra isomorphic to the matrix algebra* $\mathbf{M}_2(\mathbb{R})$.

Hint for proof. Try mapping e_1 to $\begin{bmatrix} 1 & 0 \\ 0 & -1 \end{bmatrix}$ and e_2 to $\begin{bmatrix} 0 & 1 \\ 1 & 0 \end{bmatrix}$.

Example 13.1.28. The Clifford algebra $\mathrm{Cl}(2,0)$ is *not* a $\mathbb{Z}^+$-graded algebra.

Proposition 13.1.29. *The Clifford algebra* $\mathrm{Cl}(2,0)$ *does* not *have the cancellation property.*

Hint for proof. Consider uv and uw where $u = e_2 - e_{12}$, $v = e_1 + e_2$, and $w = 1 + e_2$.

13.2. Geometric Algebra in 3-Space

Definition 13.2.1. We now give the Grassmann algebra $\bigwedge(\mathbb{R}^3) = \bigoplus_{k=0}^{3} \bigwedge^k(\mathbb{R}^3)$ a new multiplication, called the *Clifford product* (or *geometric product*). The resulting Clifford algebra will be denoted by $\mathrm{Cl}(3,0)$. Recall that the vector space $\bigwedge(\mathbb{R}^3)$ is 8-dimensional. If we let $\{e_1, e_2, e_3\}$ be the standard basis for $\mathbb{R}^3$, then

$$\bigwedge\nolimits^0(\mathbb{R}^3) = \mathrm{span}\{\mathbf{1}\},$$

$$\bigwedge\nolimits^1(\mathbb{R}^3) = \mathrm{span}\{e_1, e_2, e_3\},$$

$$\bigwedge\nolimits^2(\mathbb{R}^3) = \mathrm{span}\{e_2 \wedge e_3, e_3 \wedge e_1, e_1 \wedge e_2\}, \text{ and}$$

$$\bigwedge\nolimits^3(\mathbb{R}^3) = \mathrm{span}\{e_1 \wedge e_2 \wedge e_3\}.$$

As in $\mathrm{Cl}(2,0)$ we use the abbreviations $e_{23} = e_2 \wedge e_3$, $e_{31} = e_3 \wedge e_1$, and $e_{12} = e_1 \wedge e_2$. Additionally we let $e_{123} = e_1 \wedge e_2 \wedge e_3$. As before, members of $\bigwedge^0(\mathbb{R}^3)$ are called SCALARS. Element homogeneous of degree 1 (that is, belonging to $\bigwedge^1(\mathbb{R}^2)$) are called VECTORS (or a 1-VECTORS). Elements homogeneous of degree 2 (that is, belonging to $\bigwedge^2(\mathbb{R}^2)$) are called BIVEC-TORS. Decomposable elements of $\bigwedge^2(\mathbb{R}^2)$ are called 2-BLADES. (Thus any bivector is a sum of 2-blades.) And elements homogeneous of degree 3 (that is, scalar multiples of e_{123}) are called TRIVECTORS (or PSEUDOSCALARS, or 3-BLADES).

We start by taking scalar multiplication to be exactly as it is in the Grassmann algebra $\bigwedge(\mathbb{R}^3)$. Then we define the CLIFFORD PRODUCT or

GEOMETRIC PRODUCT of two 1-vectors, v and w, exactly as we did for $\mathrm{Cl}(2,0)$:

$$vw := \langle v, w \rangle + v \wedge w. \qquad (13.2)$$

Additionally, we set $e_1 e_2 e_3 := e_1 \wedge e_2 \wedge e_3$.

Exercise 13.2.2. Show that in order for Definition 13.2.1 to produce an algebra we need the following multiplication table to hold for the basis elements.

	1	e_1	e_2	e_3	e_{23}	e_{31}	e_{12}	e_{123}
1	1	e_1	e_2	e_3	e_{23}	e_{31}	e_{12}	e_{123}
e_1	e_1	1	e_{12}	$-e_{31}$	e_{123}	$-e_3$	e_2	e_{23}
e_2	e_2	$-e_{12}$	1	e_{23}	e_3	e_{123}	$-e_1$	e_{31}
e_3	e_3	e_{31}	$-e_{23}$	1	$-e_2$	e_1	e_{123}	e_{12}
e_{23}	e_{23}	e_{123}	$-e_3$	e_2	-1	$-e_{12}$	e_{31}	$-e_1$
e_{31}	e_{31}	e_3	e_{123}	$-e_1$	e_{12}	-1	$-e_{23}$	$-e_2$
e_{12}	e_{12}	$-e_2$	e_1	e_{123}	$-e_{31}$	e_{23}	-1	$-e_3$
e_{123}	e_{123}	e_{23}	e_{31}	e_{12}	$-e_1$	$-e_2$	$-e_3$	-1

Notation 13.2.3. The object defined in 13.2.1, the vector space $\bigwedge(\mathbb{R}^3)$ together with Clifford multiplication, turns out to be a unital algebra and is a second example of a *Clifford algebra*. We will see in the next chapter why this algebra is denoted by $\mathrm{Cl}(3,0)$.

In the preceding section we commented on the fact that, geometrically speaking, we picture a vector in the plane as an equivalence class of directed intervals (that is, directed line segments), two intervals being equivalent if they lie on parallel lines, have the same length, and point in the same direction. In a similar fashion it is helpful to have a geometrical interpretation of a 2-blade. Let us say that an *directed interval* in 3-space is an oriented parallelogram generated by a pair of vectors. Two such intervals will be said to be equivalent if they lie in parallel planes, have the same area, and have the same orientation. We may choose as a standard representative of the 2-blade $v \wedge w$ the parallelogram formed by putting the tail of v at the origin and the tail of w at the head of v. (Complete the parallelogram with by adding, successively, the vectors $-v$ and $-w$.) The standard representative of the 2-blade $w \wedge v = -v \wedge w$ is the same parallelogram with the opposite orientation: put the tail of w at the origin and the tail of v at the head of w.

Proposition 13.2.4. *The Clifford algebra* Cl(3, 0) *is indeed a unital algebra.*

Hint for proof. Consider the subalgebra of $\mathbf{M}_4(\mathbb{R})$ generated by the matrices

$$\varepsilon_1 = \begin{bmatrix} 0 & 0 & 0 & 1 \\ 0 & 0 & 1 & 0 \\ 0 & 1 & 0 & 0 \\ 1 & 0 & 0 & 0 \end{bmatrix}, \quad \varepsilon_2 = \begin{bmatrix} 0 & 0 & 1 & 0 \\ 0 & 0 & 0 & -1 \\ 1 & 0 & 0 & 0 \\ 0 & -1 & 0 & 0 \end{bmatrix}, \quad \text{and}$$

$$\varepsilon_3 = \begin{bmatrix} 1 & 0 & 0 & 0 \\ 0 & 1 & 0 & 0 \\ 0 & 0 & -1 & 0 \\ 0 & 0 & 0 & -1 \end{bmatrix}.$$

Exercise 13.2.5. The proof of the preceding proposition provides a representation of the algebra Cl(3, 0) as an 8-dimensional algebra of matrices of real numbers. Another, perhaps more interesting, representation of the same algebra is by means of the *Pauli spin matrices:*

$$\sigma_1 = \begin{bmatrix} 0 & 1 \\ 1 & 0 \end{bmatrix}, \quad \sigma_2 = \begin{bmatrix} 0 & -i \\ i & 0 \end{bmatrix}, \quad \text{and} \quad \sigma_3 = \begin{bmatrix} 1 & 0 \\ 0 & -1 \end{bmatrix}.$$

Let $\mathbb{C}(2)$ be the *real* vector space of 2×2-matrices with complex entries

(a) Show that although the *complex* vector space $M_2(\mathbb{C})$ is 4-dimensional, $\mathbb{C}(2)$ is 8-dimensional.
(b) Show that $\{I, \sigma_1, \sigma_2, \sigma_3, \sigma_2\sigma_3, \sigma_3\sigma_1, \sigma_1\sigma_2, \sigma_1\sigma_2\sigma_3\}$ is a linearly independent subset of $\mathbb{C}(2)$. (Here, I is the 2×2 identity matrix.)
(c) Find an explicit isomorphism from Cl(3, 0) to $\mathbb{C}(2)$.

As was the case in the plane we wish to assign magnitudes to 2-blades. We adopt the same convention: the magnitude of $v \wedge w$ is the area of the parallelogram generated by v and w.

Definition 13.2.6. Let v and w be 1-vectors in Cl(3, 0). Define $\|v \wedge w\| :=$ $\|v\|\|w\| \sin \theta$ where θ is the angle between v and w $(0 \leq \theta \leq \pi)$.

Proposition 13.2.7. *Let u be a unit vector in $\mathbb{R}^3$ and P be a plane through the origin perpendicular to u. Then the reflection $\dot{v}$ of a vector $v \in \mathbb{R}^3$ with respect to P is given by*

$$\dot{v} = -uvu.$$

Hint for proof. If $v_\parallel$ and $v_\perp$ are, respectively, the parallel and perpendicular components of v with respect to u, then $\dot{v} = v_\perp - v_\parallel$.

Exercise 13.2.8. Using the usual vector notation, how would you express $\dot{v}$ in the preceding proposition?

Exercise 13.2.9. Find the area of the triangle in $\mathbb{R}^3$ whose vertices are the origin, $(5, -4, 2)$, and $(1, -4, -6)$.

Example 13.2.10. To rotate a vector in 3-space by 2θ radians about an axis determined by a nonzero vector a, choose unit vectors u_1 and u_2 perpendicular to a so that the angle between u_1 and u_2 is θ radians. Let P_1 and P_2 be the planes through the origin perpendicular to u_1 and u_2, respectively. Then the map $v \mapsto \ddot{v} = R^{-1}vR$ where $R = u_1 u_2$ is the desired rotation.

Example 13.2.11. It is clear that a $90°$ rotation in the positive (counterclockwise) direction about the z-axis in $\mathbb{R}^3$ will take the point $(0, 2, 3)$ to the point $(-2, 0, 3)$. Show how the formula derived in the previous example gives this result when we choose $u_1 = e_1$, $u_2 = \frac{1}{\sqrt{2}}e_1 + \frac{1}{\sqrt{2}}e_2$, $a = (0, 0, 1)$, and $v = 2e_2 + 3e_3$.

Chapter 14

CLIFFORD ALGEBRAS

In this last chapter we will barely scratch the surface of the fascinating subject of Clifford algebras. For a more substantial introduction there are now, fortunately, many sources available. Among my favorites are [1], [7], [10], [14], [15], [24], [25], and [31].

14.1. Quadratic Forms

Definition 14.1.1. A bilinear form B on a real vector space V is SYMMETRIC if $B(u,v) = B(v,u)$ for all u, $v \in V$.

Definition 14.1.2. Let V be a real vector space. A function $Q: V \to \mathbb{R}$ is a QUADRATIC FORM if

(i) $Q(v) = Q(-v)$ for all v, and
(ii) the map $B: V \times V \to \mathbb{R}: (u,v) \mapsto Q(u+v) - Q(u) - Q(v)$ is a bilinear form.

In this case B is the bilinear form ASSOCIATED WITH the quadratic form Q. It is obviously symmetric. *Note:* In many texts $B(u,v)$ is defined to be $\frac{1}{2}[Q(u+v) - Q(u) - Q(v)]$.

Proposition 14.1.3. *Let Q be a quadratic form on a real vector space. Then $Q(0) = 0$.*

Example 14.1.4. Let B be a symmetric bilinear form on a real vector space V. If we define $Q: V \to \mathbb{R}$ by $Q(v) = B(v,v)$, then Q is a quadratic form on V.

Proposition 14.1.5. *If Q is a quadratic form on a real vector space V, then*

$$Q(u+v+w) - Q(u+v) - Q(u+w) - Q(v+w) + Q(u) + Q(v) + Q(w) = 0$$

for all u, v, $w \in V$.

Corollary 14.1.6. *If Q is a quadratic form on a real vector space V, then $Q(2v) = 4Q(v)$ for every $v \in V$.*

Hint for proof. Look at $Q(v + v - v)$.

Proposition 14.1.7. *If Q is a quadratic form on a real vector space V, then $Q(\alpha v) = \alpha^2 Q(v)$ for all $\alpha \in \mathbb{R}$ and $v \in V$.*

14.2. Definition of Clifford Algebra

Definition 14.2.1. Let V be a real vector space. A pair (U, ι), where U is a real unital algebra and $\iota \colon V \to U$ is a linear map, is UNIVERSAL OVER V if for every real unital algebra A and every linear map $f \colon V \to A$ there exists a unique unital algebra homomorphism $\widetilde{f} \colon U \to A$ such that $\widetilde{f} \circ \iota = f$.

Proposition 14.2.2. *Let V be a real vector space. If U and U' are unital algebras universal over V, then they are isomorphic.*

Example 14.2.3. For every real vector space V there is a unital algebra that is universal over V.

Hint for proof. See 8.2.2.

Definition 14.2.4. Let V be a real vector space with a quadratic form Q and A be a real unital algebra. A map $f \colon V \to A$ is a CLIFFORD MAP if

(i) f is linear, and
(ii) $\big(f(v)\big)^2 = Q(v)\mathbf{1}_A$ for every $v \in V$.

Proposition 14.2.5. *Condition (ii) in Definition 14.2.4 is equivalent to*

(ii$'$) $f(u)f(v) + f(v)f(u) = B(u,v)\mathbf{1}_A$ for all u, $v \in V$,

where B is the bilinear form associated with Q.

Definition 14.2.6. Let V be a real vector space with a quadratic form Q. The CLIFFORD ALGEBRA OVER V is a real unital algebra $\mathrm{Cl}(V, Q)$, together with a Clifford map $j \colon V \to \mathrm{Cl}(V, Q)$, that satisfies the following *universal condition:* for every real unital algebra A and every Clifford map $f \colon V \to A$, there exists a unique unital algebra homomorphism $\widehat{f} \colon \mathrm{Cl}(V, Q) \to A$ such that $\widehat{f} \circ j = f$.

Proposition 14.2.7. *Let V be a real finite dimensional vector space with a quadratic form Q. If the Clifford algebra $\mathrm{Cl}(V, Q)$ exists, then it is unique up to isomorphism.*

Example 14.2.8. For every real finite dimensional vector space V with a quadratic form Q the Clifford algebra $\mathrm{Cl}(V, Q)$ exists.

Hint for proof. Try $\mathcal{T}(V)/J$ where J is the ideal in $\mathcal{T}(V)$ generated by elements of the form $v \otimes v - Q(v)\mathbf{1}_{\mathcal{T}(V)}$ where $v \in V$.

14.3. Orthogonality with Respect to Bilinear Forms

Definition 14.3.1. Let B be a symmetric bilinear form on a real vector space V. Vectors v and w in V are ORTHOGONAL, in which case we write $v \perp w$, if $B(v, w) = 0$. The KERNEL of B is the set of all $k \in V$ such that $k \perp v$ for every $v \in V$. The bilinear form is NONDEGENERATE if its kernel is $\{\mathbf{0}\}$.

Exercise 14.3.2. One often sees the claim that "the classification of Clifford algebras amounts to classifying vector spaces with quadratic forms". Explain precisely what is meant by this assertion.

Proposition 14.3.3. *Let B be a symmetric bilinear form on a real finite dimensional vector space V. Suppose that V is an orthogonal direct sum $V_1 \oplus \cdots \oplus V_n$ of subspaces. Then B is nondegenerate if and only if the restriction of B to V_k is nondegenerate for each k. In fact, if we denote by B_j the restriction of B to V_j, then $\ker B = \ker B_1 \oplus \cdots \oplus \ker B_n$.*

Proposition 14.3.4. *Let Q be a quadratic form on a real finite dimensional vector space V and let $\{e_1, \ldots, e_n\}$ be a basis for V that is orthogonal with respect to the bilinear form B associated with Q. If $Q(e_k)$ is nonzero for $1 \leq k \leq p$ and $Q(e_k) = 0$ for $p < k \leq n$, then the kernel of B is the span of $\{e_{p+1}, \ldots, e_n\}$.*

Proposition 14.3.5. *Let Q be a quadratic form on a real finite dimensional vector space V. If $\dim V = n$, then $\dim \mathrm{Cl}(V, Q) = 2^n$.*

Exercise 14.3.6. Let V be a real finite dimensional vector space and let Q be the quadratic form that is identically zero on V. Identify $\mathrm{Cl}(V, Q)$.

14.4. Examples of Clifford Algebras

Definition 14.4.1. Let V be a finite dimensional real vector space and B be a symmetric bilinear form on V. An ordered basis $E = (e^1, \ldots, e^n)$ for V is B-ORTHONORMAL if

(a) $B(e^i, e^j) = 0$ whenever $i \neq j$ and
(b) for each $i \in \mathbb{N}_n$ the number $B(e^i, e^i)$ is -1 or $+1$ or 0.

Theorem 14.4.2. *If V is a finite dimensional real vector space and B is a symmetric bilinear form on V, then V has a B-orthonormal basis.*

Proof. See [4], Chapter 1, Theorem 7.6.

Convention 14.4.3. Let V be a finite dimensional real vector space, let B be a symmetric bilinear form on V, and let Q be the quadratic form associated with B. Let us agree that whenever $E = (e^1, \ldots, e^n)$ is an ordered B-orthonormal basis for V, we order the basis elements in such a way that for some positive integers p and q

$$Q(e^i) = \begin{cases} 1, & \text{if } 1 \leq i \leq p; \\ -1, & \text{if } p + 1 \leq i \leq p + q; \\ 0, & \text{if } p + q + 1 \leq i \leq n. \end{cases}$$

Theorem 14.4.4. *Let V be a finite dimensional real vector space, let B be a symmetric bilinear form on V, and let Q be the quadratic form associated with B. Then there exist $p, q \in \mathbb{Z}^+$ such that if $E = (e^1, \ldots, e^n)$ is a B-orthonormal basis for V and $v = \sum v_k e^k$, then*

$$Q(v) = \sum_{k=1}^{p} v_k{}^2 - \sum_{k=p+1}^{p+q} v_k{}^2.$$

Proof. See [4], Chapter 1, Theorem 7.11.

Notation 14.4.5. If (V, Q) is a finite dimensional real vector space V with a nondegenerate quadratic form Q, we often denote the Clifford algebra $\mathrm{Cl}(V, Q)$ by $\mathrm{Cl}(p, q)$ where p and q are as in Theorem 14.4.4.

Proposition 14.4.6. *Let* $f\colon (V, Q) \to (W, R)$ *be a linear map between finite dimensional real vector spaces with quadratic forms. If* $R(f(v)) = Q(v)$ *for every* $v \in V$ *we say that* f *is an* ISOMETRY.

(a) *If* f *is such a linear isometry, then there exists a unique unital algebra homomorphism* $\mathrm{Cl}(f)\colon \mathrm{Cl}(V, Q) \to \mathrm{Cl}(W, R)$ *such that* $\mathrm{Cl}(f)(v) = f(v)$ *for every* $v \in V$.

(b) *The pair of mappings* Cl *described above is a covariant functor from the category of vector spaces with quadratic forms and linear isometries to the category of Clifford algebras and unital algebra homomorphisms.*

(c) *If a linear isometry* f *is an isomorphism, then* $\mathrm{Cl}(f)$ *is an algebra isomorphism.*

Proposition 14.4.7. *Let* V *be a real finite dimensional vector space, Q be a quadratic form on V, and $A = \mathrm{Cl}(V, Q)$ be the associated Clifford algebra.*

(a) *The map* $f\colon V \to V\colon v \mapsto -v$ *is a linear isometry.*

(b) *If* $\omega = \mathrm{Cl}(f)$, *then* $\omega^2 = \mathrm{id}$.

(c) *Let* $A_0 = \{a \in A\colon \omega(a) = a\}$ *and* $A_1 = \{a \in A\colon \omega(a) = -a\}$. *Then* $A = A_0 \oplus A_1$.

(d) *If* $i, j \in \{0, 1\}$, *then* $A_i A_j \subseteq A_{i+j}$ *(where $i + j$ indicates addition modulo 2). This says that a Clifford algebra is a* $\mathbb{Z}_2$-GRADED *(or* $\mathbb{Z}/2\mathbb{Z}$-GRADED*) algebra.*

Hint for proof of (c). If $a \in A$, let $a_0 = \frac{1}{2}(a + \omega(a))$.

Example 14.4.8. Let $V = \mathbb{R}$ and $Q(v) = v^2$ for every $v \in V$. Then the Clifford algebra $\mathrm{Cl}(1, 0)$ associated with $(\mathbb{R}, Q)$ is isomorphic to $\mathbb{R} \oplus \mathbb{R}$.

Hint for proof. Let $f\colon \mathbb{R} \to \mathrm{Cl}(1, 0)$ be the Clifford map associated with Q and let $e := f(1)$. Try the function $\phi\colon \mathrm{Cl}(1, 0) \to \mathbb{R} \oplus \mathbb{R}$ under which $1 \mapsto (1, 1)$ and $e \mapsto (-1, 1)$.

Example 14.4.9. Let $V = \mathbb{R}$ and $Q(v) = -v^2$ for every $v \in V$.

(a) The algebra $\mathrm{Cl}(0, 1)$ is isomorphic to $\mathbb{C}$.

(b) The algebra $\mathrm{Cl}(0, 1)$ can be represented as a subalgebra of $M_2(\mathbb{R})$.

Hint for proof. For part (a) keep in mind that as a real algebra $\mathbb{C}$ is 2-dimensional. For part (b) consider the matrices $\begin{bmatrix} 1 & 0 \\ 0 & 1 \end{bmatrix}$ and $\begin{bmatrix} 0 & -1 \\ 1 & 0 \end{bmatrix}$.

Example 14.4.10. Let $V = \mathbb{R}^2$ and $Q(v) = v_1{}^2 + v_2{}^2$ for every $v \in V$. Then $\mathrm{Cl}(2, 0) \cong M_2(\mathbb{R})$. (See 13.1.2 and 13.1.4.)

Example 14.4.11. Let $V = \mathbb{R}^3$ and $Q(v) = v_1{}^2 + v_2{}^2 + v_3{}^2$ for every $v \in V$. Then $\mathrm{Cl}(3,0)$ is isomorphic to both a subalgebra of $\mathbf{M}_4$ and the algebra $\mathbb{C}(2)$ generated by the *Pauli spin matrices*. (See 13.2.4 and 13.2.5.)

In 1833, when William Rowan Hamilton was still in his twenties, he presented a paper to the *Royal Irish Academy* in which he explained how to represent complex numbers as pairs of real numbers. He was fascinated by the connections this produced between complex arithmetic and the geometry of the plane. For more than a decade he searched, sometimes quite obsessively, for a 3-dimensional algebraic analog that would facilitate studies in the geometry of 3-space. (In modern language, he was seeking a 3-dimensional normed division algebra. As it turns out, there aren't any!) On October 16, 1843, while walking with his wife along the Royal Canal in Dublin on his way to chair a meeting of the *Royal Irish Academy*, he suddenly understood that to manage triples of real numbers algebraically, he would need to add, in some sense, a fourth dimension. The objects he could work with were of the form $a + \mathbf{i}b + \mathbf{j}c + \mathbf{k}d$ where a, b, c, and d are real numbers. In this flash of inspiration he saw the necessary rules for multiplication of these objects. Although, as far as we know, he had no previous history as a graffiti artist, he stopped with his wife to carve into the stones of Brougham Bridge, which they were passing, the key to the multiplication of these new objects, quaternions:

$$\mathbf{i}^2 = \mathbf{j}^2 = \mathbf{k}^2 = \mathbf{i}\mathbf{j}\mathbf{k} = -1.$$

Exercise 14.4.12. Derive from the preceding equations the multiplication table for quaternions.

Example 14.4.13. Let $V = \mathbb{R}^2$ and $Q(v) = -v_1^2 - v_2^2$ for every $v \in V$.

(a) The algebra $\mathrm{Cl}(0,2)$ is isomorphic to the algebra $\mathbb{H}$ of quaternions.
(b) The algebra $\mathrm{Cl}(0,2)$ can be represented as a subalgebra of $M_4(\mathbb{R})$.

Hint for proof. For part (b) consider the matrices $\begin{bmatrix} 0 & -1 & 0 & 0 \\ 1 & 0 & 0 & 0 \\ 0 & 0 & 0 & -1 \\ 0 & 0 & 1 & 0 \end{bmatrix}$ and

$\begin{bmatrix} 0 & 0 & -1 & 0 \\ 0 & 0 & 0 & 1 \\ 1 & 0 & 0 & 0 \\ 0 & -1 & 0 & 0 \end{bmatrix}.$

Example 14.4.14. The Clifford algebra $Cl(1,1)$ is isomorphic to $\mathbf{M}_2(\mathbb{R})$.

Hint for proof. Consider the matrices $\begin{bmatrix} 1 & 0 \\ 0 & -1 \end{bmatrix}$ and $\begin{bmatrix} 0 & 1 \\ -1 & 0 \end{bmatrix}$.

Exercise 14.4.15. Take a look at the web page [23] written by Pertti Lounesto.

Exercise 14.4.16. The Clifford algebra $Cl(3,1)$ (*Minkowski space-time algebra*) is isomorphic to $M_4(\mathbb{R})$.

Hint for proof. Exercise 14.4.15.

BIBLIOGRAPHY

1. Rafal Abłamowicz and Garret Sobczyk (eds.), *Lectures on Clifford (Geometric) Algebras and Applications*, Birkhäuser, Boston, 2004. 125
2. Ralph Abraham, Jerrold E. Marsden, and Tudor Ratiu, *Manifolds, Tensor Analysis, and Applications*, Addison-Wesley, Reading, MA, 1983. 114, 115, 116
3. Richard L. Bishop and Richard J. Crittenden, *Geometry of Manifolds*, Academic Press, New York, 1964. 83, 99
4. William C. Brown, *A Second Course in Linear Algebra*, John Wiley, New York, 1988. 127
5. Stewart S. Cairns, A simple triangulation method for smooth manifolds, *Bull. Amer. Math. Soc.* **67** (1961), 389–390. 110
6. P. M. Cohn, *Basic Algebra: Groups, Rings and Fields*, Springer, London, 2003. 9
7. John S. Denker, *An Introduction to Clifford Algebra*, 2006, http://www.av8n.com/physics/clifford-intro.htm. 125
8. J. Dieudonné, *Treatise on Analysis, Volumes I and II*, Academic Press, New York, 1969, 1970. 67
9. Manfredo P. do Carmo, *Differential Forms and Applications*, Springer-Verlag, Berlin, 1994. 116
10. Chris Doran and Anthony Lasenby, *Geometric Algebra for Physicists*, Cambridge University Press, Cambridge, 2003. 125
11. John M. Erdman, *A Companion to Real Analysis*, 2007, http://web.pdx.edu/~erdman/CRA/CRAlicensepage.html. 67
12. _____, *A Problems Based Course in Advanced Calculus*, American Mathematical Society, Providence, Rhode Island, 2018. 67, 86, 91
13. Douglas R. Farenick, *Algebras of Linear Transformations*, Springer-Verlag, New York, 2001. 48, 50
14. D. J. H. Garling, *Clifford Algebras: An Introduction*, Cambridge University Press, Cambridge, 2011. 125

15. David Hestenes and Garret Sobczyk, *Clifford Algebra to Geometric Calculus*, D. Reidel, Dordrecht, 1984. 125

16. Kenneth Hoffman and Ray Kunze, *Linear Algebra*, second ed., Prentice Hall, Englewood Cliffs,N.J., 1971. 52, 65

17. S. T. Hu, *Differentiable Manifolds*, Holt, Rinehart, and Winston, New York, 1969. 110

18. Thomas W. Hungerford, *Algebra*, Springer-Verlag, New York, 1974. 9

19. Saunders Mac Lane and Garrett Birkhoff, *Algebra*, Macmillan, New York, 1967. 9

20. Serge Lang, *Fundamentals of Differential Geometry*, Springer Verlag, New York, 1999. 115

21. John M. Lee, *Introduction to Smooth Manifolds*, Springer, New York, 2003. 83, 99, 110

22. Lynn H. Loomis and Shlomo Sternberg, *Advanced calculus*, Jones and Bartlett, Boston, 1990. 67

23. Pertti Lounesto, *Counterexamples in Clifford Algebras*, 1997/2002, http://users.tkk.fi/~ppuska/mirror/Lounesto/counterexamples.htm. 128

24. _____, *Clifford Algebras and Spinors*, second ed., Cambridge University Press, Cambridge, 2001. 125

25. Douglas Lundholm and Lars Svensson, *Clifford algebra, geometric algebra, and applications*, 2009, http://arxiv.org/abs/0907.5356v1. 125

26. Ib Madsen and Jørgen Tornehave, *From Calculus to Cohomology: de Rham cohomology and characteristic classes*, Cambridge University Press, Cambridge, 1997. 115

27. Theodore W. Palmer, *Banach Algebras and the General Theory of *-Algebras I–II*, Cambridge University Press, Cambridge, 1994/2001. 9

28. Charles C. Pinter, *A Book of Abstract Algebra*, second ed., McGraw-Hill, New York, 1990. 73

29. Steven Roman, *Advanced Linear Algebra*, second ed., Springer-Verlag, New York, 2005. 19, 65, 76

30. I. M. Singer and J. A. Thorpe, *Lecture Notes on Elementary Topology and Geometry*, Springer Verlag, New York, 1967. 110

31. John Snygg, *A New Approach to Differential Geometry using Clifford's Geometric Algebra*, Springer, New York, 2012. 125

32. Gerard Walschap, *Metric Structures in Differential Geometry*, Springer-Verlag, New York, 2004. 99

33. Frank W. Warner, *Foundations of Differentiable Manifolds and Lie Groups*, Springer Verlag, New York, 1983. 110, 115

INDEX